Metal and Metal Oxide Nanoparticles: Design, Characterization, and Biomedical Applications

Metal and Metal Oxide Nanoparticles: Design, Characterization, and Biomedical Applications

Editor

Tania Limongi

MDPI • Basel • Beijing • Wuhan • Barcelona • Belgrade • Manchester • Tokyo • Cluj • Tianjin

Editor
Tania Limongi
DISAT
Politecnico di Torino
Corso Duca Degli Abruzzi 24
Italy

Editorial Office
MDPI
St. Alban-Anlage 66
4052 Basel, Switzerland

This is a reprint of articles from the Special Issue published online in the open access journal *Materials* (ISSN 1996-1944) (available at: www.mdpi.com/journal/materials/special_issues/Metal_Metal_Oxide_Nanopart_Des_Charact_Biomed_Appl).

For citation purposes, cite each article independently as indicated on the article page online and as indicated below:

LastName, A.A.; LastName, B.B.; LastName, C.C. Article Title. *Journal Name* **Year**, *Volume Number*, Page Range.

ISBN 978-3-0365-3169-4 (Hbk)
ISBN 978-3-0365-3168-7 (PDF)

Contents

About the Editor

Tania Limongi

Tania Limongi is an Assistant Professor in Italy in the Department of Applied Science and Technology, of the Politecnico of Torino. She received her Master Degree in 1999 in Biological Sciences at the University of L'Aquila and her Ph.D. in Microsystem Engineering at the University of Rome Tor Vergata in 2004. Dr Limongi's research interests are primarily in Applied Biology, Biomedical Engineering, Biochemistry, and Molecular Physics. Her current research focuses on the development of effective drug delivery solution and hybrid nanoparticles for combined therapeutic and diagnostic applications, on the fabrication of biocompatible and biodegradable porous scaffolds for tissue engineering application and on the optimization of new effective high resolution single molecule detection approaches. She published more than 70 papers in reputed journals, she holds 3 patents, and she is reviewer, section editor and guest editor for several peer review journals.

Preface to "Metal and Metal Oxide Nanoparticles: Design, Characterization, and Biomedical Applications"

The current Special Issue entitled "Metal and Metal Oxide Nanoparticles: Design, Characterization, and Biomedical Applications"aims to present contributions from all scientists producing and/or applying metal and metal oxide nanoparticles in a diagnostic, therapeutic or theranostics context.

Developing new materials is usually a time-demanding and meticulous process, but at the same time, it is one of the more promising solutions to obtain a cleaner, safer, and smart future. In more detail, when referring to nanomaterials, an increasingly successfully tool of nanotechnologies, nanoparticles are categorized as materials in which at least one dimension is less than 100 nm in diameter. Among the various nanoparticle categories, metal and metal oxide nanoparticles stand out as emerging nanotechnological solutions for a wide range of biological and medical physio/pathological open questions. This Special Issue covers the fundamental science, design, characterization, and biomedical applications of metal and metal oxide nanomaterials and the five papers here presented embrace all the aspects determining the performance of these systems, ranging from their synthesis, design, chemical, physical, and biological functionalization, to their characterization and successful applications.

Metal-based nanoparticles (MNPs) include metal NPs, metal oxide NPs, quantum dots (QDs) and magnetic NPs and, thanks to their chemical physical properties, have gained much traction for their functional use in biomedicine.

Tania Limongi
Editor

materials

Editorial

Special Issue "Metal and Metal Oxide Nanoparticles: Design, Characterization, and Biomedical Applications"

Tania Limongi

Department of Applied Science and Technology, Politecnico di Torino, Corso Duca degli Abruzzi 24, 10129 Turin, Italy; tania.limongi@polito.it

The current Special Issue entitled "Metal and Metal Oxide Nanoparticles: Design, Characterization, and Biomedical Applications" aims to present contributions from all scientists producing and/or applying metal and metal oxide nanoparticles in a diagnostic, therapeutic or theranostics context.

Developing new materials is usually a time-demanding and meticulous process, but at the same time, it is one of the more promising solutions to obtain a cleaner, safer, and smart future. In more detail, when referring to nanomaterials, an increasingly successfully tool of nanotechnologies, nanoparticles are categorized as materials in which at least one dimension is less than 100 nm in diameter. Among the various nanoparticle categories, metal and metal oxide nanoparticles stand out as emerging nanotechnological solutions for a wide range of biological and medical physio/pathological open questions [1–3]. This Special Issue covers the fundamental science, design, characterization, and biomedical applications of metal and metal oxide nanomaterials and the five papers here presented embrace all the aspects determining the performance of these systems, ranging from their synthesis, design, chemical, physical, and biological functionalization, to their characterization and successful applications.

Metal-based nanoparticles (MNPs) include metal NPs, metal oxide NPs, quantum dots (QDs) and magnetic NPs and, thanks to their chemical physical properties, have gained much traction for their functional use in biomedicine.

In the last decades, many authors have described how the use of MNPs in stimuli-responsive systems design encourages their clinical translatability, improving also their preclinical investigation [4–8]. In their review, Caneparo et al. highlighted how reactive oxygen species (ROS) generation could be considered an effective nanotechnology tool for alternative therapies, such as photodynamic therapy (PDT), high-intensity focused ultrasound therapy (HIFU), photothermal therapy (PPT) and sonodynamic therapy (SDT) [9].

Laurenti et al. explored the use of a new composite material based on the incorporation of mesoporous flower-like ZnO micropowders into polyHEMA and poly(HEMA-co-AA) hydrogels for drug eluting stent applications. Their release study pointed out that the poly(HEMA-co-AA)@ZnO_0.1% formulation satisfied the multifunctional requirements needed in the field of ureteral stent applications such as antibacterial effects, drug elution and biodegradability [10].

Bartoli et al., summarizing in their review the complex chemical behavior of bismuth during the transformation of its compounds to oxide and bismuth oxide-phase transitions, reported the main achievements reached by using nanostructured bismuth oxide and related nanomaterials for the design and the application of new diagnostic and theranostic tools [11].

Svensson et al. in their communication, reported a proof of principle for the self-assembly of titania nanoshells from anisotropically functionalized titania NPs successfully utilized for fast-release drug delivery application [12].

To conclude, Mylkie et al. described the efficacy of biocompatible natural polymer-coated magnetite NPs for protein immobilization, highlighting their application in biomedicine [13].

Citation: Limongi, T. Special Issue "Metal and Metal Oxide Nanoparticles: Design, Characterization, and Biomedical Applications". *Materials* **2021**, *14*, 7287. https://doi.org/10.3390/ma14237287

Received: 21 November 2021
Accepted: 26 November 2021
Published: 28 November 2021

Funding: This research received no external funding.

Acknowledgments: The Guest Editor would like to thank all the authors from all over the world (Sweden, USA, Poland and Italy), who contributed with their valuable works to the accomplishment of the Special Issue. Special thanks are due to the reviewers for their constructive comments and thoughtful suggestions and to the *Materials* Editorial Office, particularly to Yulia Zhao for her kind assistance.

Conflicts of Interest: The author declares no conflict of interest.

References

1. Limongi, T.; Susa, F.; Cauda, V. Nanoparticles for hematologic diseases detection and treatment. *Hematol. Med. Oncol.* **2019**, *4*, 1000183. [CrossRef] [PubMed]
2. Limongi, T.; Canta, M.; Racca, L.; Ancona, A.; Tritta, S.; Vighetto, V.; Cauda, V. Improving dispersal of therapeutic nanoparticles in the human body. *Nanomedicine* **2019**, *14*, 797–801. [CrossRef] [PubMed]
3. Dadfar, S.M.; Roemhild, K.; Drude, N.I.; von Stillfried, S.; Knüchel, R.; Kiessling, F.; Lammers, T. Iron oxide nanoparticles: Diagnostic, therapeutic and theranostic applications. *Adv. Drug Deliv. Rev.* **2019**, *138*, 302–325. [CrossRef] [PubMed]
4. Chelminiak-Dudkiewicz, D.; Rybczynski, P.; Smolarkiewicz-Wyczachowski, A.; Mlynarczyk, D.T.; Wegrzynowska-Drzymalska, K.; Ilnicka, A.; Goslinski, T.; Marszałł, M.P.; Ziegler-Borowska, M. Photosensitizing potential of tailored magnetite hybrid nanoparticles functionalized with levan and zinc (II) phthalocyanine. *Appl. Surf. Sci.* **2020**, *524*, 146602. [CrossRef] [PubMed]
5. Marino, A.; Battaglini, M.; De Pasquale, D.; Degl'Innocenti, A.; Ciofani, G. Ultrasound-Activated Piezoelectric Nanoparticles Inhibit Proliferation of Breast Cancer Cells. *Sci. Rep.* **2018**, *8*, 6257. [CrossRef] [PubMed]
6. Foglietta, F.; Serpe, L.; Canaparo, R. The Effective Combination between 3D Cancer Models and Stimuli-Responsive Nanoscale Drug Delivery Systems. *Cells* **2021**, *10*, 3295. [CrossRef]
7. Petruleviciene, M.; Juodkazyte, J.; Parvin, M.; Tereshchenko, A.; Ramanavicius, S.; Karpicz, R.; Samukaite-Bubniene, U.; Ramanavicius, A. Tuning the Photo-Luminescence Properties of WO3 Layers by the Adjustment of Layer Formation Conditions. *Materials* **2020**, *13*, 2814. [CrossRef] [PubMed]
8. Racca, L.; Limongi, T.; Vighetto, V.; Dumontel, B.; Ancona, A.; Canta, M.; Canavese, G.; Garino, N.; Cauda, V. Zinc Oxide Nanocrystals and High-Energy Shock Waves: A New Synergy for the Treatment of Cancer Cells. *Front. Bioeng. Biotechnol.* **2020**, *8*, 577. [CrossRef] [PubMed]
9. Canaparo, R.; Foglietta, F.; Limongi, T.; Serpe, L. Biomedical Applications of Reactive Oxygen Species Generation by Metal Nanoparticles. *Materials* **2021**, *14*, 53. [CrossRef] [PubMed]
10. Laurenti, M.; Grochowicz, M.; Dragoni, E.; Carofiglio, M.; Limongi, T.; Cauda, V. Biodegradable and Drug-Eluting Inorganic Composites Based on Mesoporous Zinc Oxide for Urinary Stent Applications. *Materials* **2020**, *13*, 3821. [CrossRef] [PubMed]
11. Bartoli, M.; Jagdale, P.; Tagliaferro, A. A Short Review on Biomedical Applications of Nanostructured Bismuth Oxide and Related Nanomaterials. *Materials* **2020**, *13*, 5234. [CrossRef] [PubMed]
12. Svensson, F.G.; Seisenbaeva, G.A.; Kotov, N.A.; Kessler, V.G. Self-Assembly of Asymmetrically Functionalized Titania Nanoparticles into Nanoshells. *Materials* **2020**, *13*, 4856. [CrossRef] [PubMed]
13. Mylkie, K.; Nowak, P.; Rybczynski, P.; Ziegler-Borowska, M. Polymer-Coated Magnetite Nanoparticles for Protein Immobilization. *Materials* **2021**, *14*, 248. [CrossRef] [PubMed]

materials

Review

A Short Review on Biomedical Applications of Nanostructured Bismuth Oxide and Related Nanomaterials

Mattia Bartoli [1,2], Pravin Jagdale [3] and Alberto Tagliaferro [1,2,*]

1 Department of Applied Science and Technology, Politecnico di Torino, Corso Duca degli Abruzzi 24, 10129 Turin, Italy; mattia.bartoli@polito.it
2 Italian Institute of Technology, Via Livorno 60, 10144 Torino, Italy
3 Consorzio Interuniversitario Nazionale per la Scienza e Tecnologia dei Materiali (INSTM), Via G. Giusti 9, 50121 Florence, Italy; pravin.jagdale@iit.it
* Correspondence: alberto.tagliaferro@polito.it; Tel.: +39-011-0907347

Received: 28 October 2020; Accepted: 18 November 2020; Published: 19 November 2020

Abstract: In this review, we reported the main achievements reached by using bismuth oxides and related materials for biological applications. We overviewed the complex chemical behavior of bismuth during the transformation of its compounds to oxide and bismuth oxide phase transitions. Afterward, we summarized the more relevant studies regrouped into three categories based on the use of bismuth species: (i) active drugs, (ii) diagnostic and (iii) theragnostic. We hope to provide a complete overview of the great potential of bismuth oxides in biological environments.

Keywords: bismuth oxide; nanoparticles; radiopacity; chemotherapy; theragnostic

1. Introduction

Recently, the concerns for human healthiness have risen day by day. Pharma industries and the research community have committed themselves to improve both the knowledge and on-field real application of newly designed drugs and protocols. Despite the numerous available established treatments, there is still an urgent need to develop new and innovative technologies that could help to delineate tumor margins, identify residual tumor cells and eventually eliminate them [1–3].

Furthermore, the development of new antimicrobial agents able to overcome antibiotic resistance rising has become critical [4].

All of these issues have been deeply investigated by nanoscience and material technology. Among all the promising nanostructured or nanosized materials, bismuth-based ones, although rather neglected, are full of unexpressed potential [5].

Bismuth has been used in plenty of applications since the 19th century for the treatment of bacterial infections but its use slowed down in the middle of the 20th century after the reversible bismuth encephalopathy occurred in France and Australia [6,7]. Nonetheless, bismuth-based medical formulations are still being used for stomach issues treatments as bismuth subsalicylate [8,9], colloidal bismuth subcitrate [10–12] or bioactive conjugated as ranitidine bismuth citrate [13].

Organo-bismuth derivatives are not the only species of great interest as nanosized oxides and related materials have raised more and more interest in the scientific community due to their cost-effective fabrication processes [14,15], high stability [16] and versatility in terms of morphology [17,18]. Furthermore, the high atomic number of bismuth brings about a high energy radiation attenuation larger than that of lead at an almost negligible risk of toxicity [19]. The combinations of bismuth properties represent a unique chance to exploit singularly or simultaneously cytotoxicity and diagnostic effects. In this paper, we overview the literature providing a synthetic although comprehensive discussion on the main achievements reached by using bismuth oxide and related materials for biological applications. After presenting the main chemical and physical properties of bismuth, we regroup the studies in three main categories dealing respectively with (i) biological activity, (ii) use for the production of biomaterials and diagnostic agents and (iii) production of theragnostic platforms.

2. Bismuth Oxide and Related Materials: Productive Strategies

Nowadays, bismuth is mainly produced as a side product of lead streams and could be isolated through the Betterton–Kroll process [20] or through an electrochemical procedure known as Betts electrolytic process [21]. It is obtained in a highly purified form for those applications where it is used as a replacement for lead [22].

Commonly, bismuth is used in form of halide, oxo-halide, nitrate and oxides derivatives. Bismuth halides (BiX_3, X = F, Cl, Br, I) are generally prepared by treating bismuth oxide in a watery medium by adding the specific HX acid. Bismuth trihalides are bipyramidal molecular species in the gas phase with angle X-Bi-X in the range 96–100° [23]. In the solid phase, they show a variety of different structures based on the halogen present in the crystals. BiF3 shows a pseudo-ionic structure with tricapped trigonal prismatic motive where bismuth atoms are surrounded by nine fluoride atoms, while the other halides show bicapped trigonal prism crystals. Bismuth oxide halides (BiOX) are formed by partial hydrolysis of bismuth halides. BiOF and BiOI can also be made by heating the corresponding halides in the air. BiOX have complex layer lattice structures [24] and, when heated up to 600 °C, BiOCl or BiOBr are decomposed by forming $Bi_{24}0_{31}X_{10}$ [25].

Moving on, bismuth can easily be produced as bismuth nitrate. Firstly, it is recovered as $Bi(NO_3)_3{\cdot}5H_2O$ through crystallization after hydrolysis of Bi_2O_3 by using concentrated nitric acid. If a diluted acid is used is possible to recover the basic salt $BiO(NO_3)$. $BiO(NO_3)$ could be also produced by precipitation treating $Bi(NO_3)_3{\cdot}5H_2O$ at 150 °C with butanol forming road-like structures as reported by Liu et al. [26]. As clearly enlighten by Briand and Burford [27] the hydrolysis of $Bi(NO_3)_3{\cdot}5H_2O$ could lead to a plethora of different compounds. Furthermore, several attempts were reported in the literature [28–32] pursuing the thermal oxodehydration of $Bi(NO_3)_3{\cdot}5H_2O$ with the formation of a series of complex species as summarized in Figure 1.

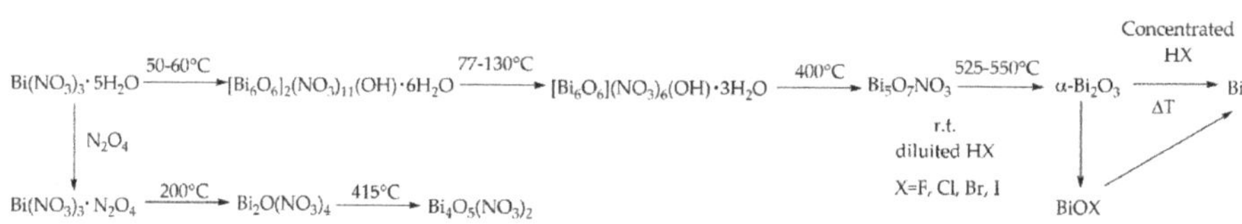

Figure 1. Comprehensive scheme of chemical evolution of $Bi(NO_3)_3{\cdot}5\,H_2O$.

An interesting study was reported by Tanveer et al. [17] about the transition from $Bi(NO_3)_3{\cdot}5H_2O$ to $Bi_5O_7NO_3$ showing how it is possible to isolate a species of $Bi_5O_7NO_3$ tailored on the surface with β-Bi_2O_3.

Bismuth oxides are the other deeply studied class of bismuth compounds and they present four different phases [33] as reported in Figure 2.

At room temperature, monoclinic α-Bi_2O_3 is the common stable phase with a polymeric-distorted layered structure composed of pentacoordinate bismuth atoms enclosed into pseudo-octahedral units. At a temperature higher than 710 °C, α phase is converted into the cubic δ phase that has a defective structure with random oxygen vacancies [34]. The β phase and several oxygen-rich forms are closely related to the δ phase. In particular, the vacancy structures of highly defected bismuth oxides some sites filled with O^{-2} together with Bi(III) and Bi (V) sites. Bismuth oxide γ-phase shows also a cubic structure but it is highly unstable and hard to synthesize without supporting it onto other oxides or metallic species [35]. The other two polymorphic metastable bismuth oxide phases are known as the ω phase stable at temperatures higher than 800 °C [36] and the ε phase isolated in 2006 by Cornei and co-workers [37].

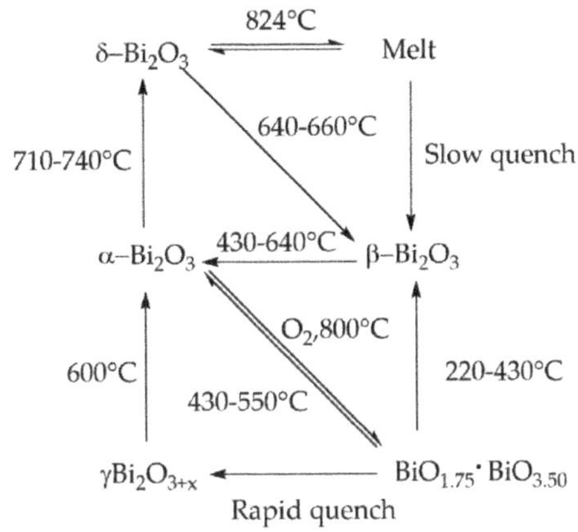

Figure 2. Scheme of phase transition of bismuth oxide.

Bismuth(V) oxides are less stable than Bi(III) but several studies reported their preparation as lithium [38] or sodium [39] salt derivatives.

Bismuth derivatives were also studied for the production of colloidal phases. Kiran et al. [40] synthesized a bismuth-substituted cobalt ferrite with a nominal formula $CoFe_2{-}0.1Bi_{0.1}O_4$ quite active for the reduction of 4-nitrophenol to 4-aminophenol in a watery solution of sodium borhydride. Metal bismuth nanoparticles were produced by Petsom et al. [41] showing that the size of the nanoparticles can be tuned by adding different amounts of ionic and non-ionic surfactants.

Furthermore, several organometallic species of bismuth such as subgallate [42] and subsalicylate [8] have found use in medical applications that will be more thoroughly discussed in the next sections and briefly summarized in Table 1.

Table 1. Summary of main properties of bismuth and related compounds.

Bismuth Species	Advantages	Issues
Metallic bismuth	■ Easy to synthesized ■ High size control ■ Highest concentration of radiopaque atoms	■ High cytotoxicity for low average size particles ■ Only spherical shaped ■ Neat surfaces without any functional groups
Organometallic bismuth	■ Hydrosoluble ■ High cellular uptaking ■ High stability	■ Low concentration of radiopaque atoms ■ Fast excretion ■ Could trespass the hematoencephalic barrier [43]
Bismuth nitrates	■ High shape tunability ■ Highly tailoring surface	■ Fast hydrolysis in watery phase under mild conditions ■ Difficult to isolated as pure compounds ■ Difficult to predict the correct active species

Table 1. *Cont.*

Bismuth Species	Advantages	Issues
Bismuth halide and oxohalides	■ Easily synthesizable ■ Photocatalytic activity	■ Oxidizable ■ Hygroscopic ■ Highly acidic
Bismuth oxides	■ Highly stable ■ Easy to synthesized ■ High size control ■ Poor cytotoxicity ■ Good cellular uptake ■ Photocatalytic activity	■ Highly hydrophobic ■ Phase impurities ■ Surface defects

3. Bismuth Oxide and Related Materials for Biological Applications

3.1. *Bismuth Based Nanomaterials as Biological Active Drugs*

The first and main point to clarify is about the interaction between bismuth-based materials and living organisms. In 1989, Slikkerveer et al. [44] reported a very comprehensive overview of the toxicity of bismuth species. As clearly emerged, the oral intake of bismuth compounds leads to a significant increase in blood concentrations of bismuth [45] and the amount rose rapidly up to 380 µmL/g [46]. Gavey et al. [47] show how the uptake could be magnified by bismuth citrate soluble species or by the simultaneous administration of cysteine [48]. Lechat et al. [49] reported a study about the administration of bismuth subnitrate showing how poorly or watery insoluble bismuth species decrement the organism uptake. As reported by several studies run on rats using $BiCl_3$ [50,51], bismuth binds to high molecular weight metallothionein protein close to those that bind copper cations [52]. Bismuth is excreted by both urine and feces but rats retain up to 10 wt.% of the dose administrated even after 90 days [53].

The in vivo tests suggest that bismuth salts or organometallic derivatives could lead to bioaccumulation and encephalopathy [54]. Stephens et al. [55] used homo- and heteroleptic bismuth(III) thiolates to prove that the bismuth complex surrounding drives the antimicrobial activity of the bismuth species.

Luo et al. [56] reported a study about the in vivo cytotoxicity of different tailored bismuth nanoparticles. The authors produced metallic bare bismuth nanoparticles and tailored their surface with amines, poly(ethyleneglycol), neat and amine tailored silica. The study was run by using HeLa and MG-63 cell lines showing cytotoxicity of bismuth species higher for the HeLa. The authors reported that the non-toxic concentration of bare nanoparticles was attested to 0.5 nM while they induce cellular death at a concentration of up to 50 nM. The functionalization decreased the cytotoxicity of the bare nanoparticles that are more toxic than the other (bare > amine-terminated > silica-coated > poly(ethylene glycol) coated). Coating increased the stability of bismuth inside the cells but decreased its ability to induce oxidative stress.

Abudayyak et al. [57] studied bismuth oxide nanoparticles like the ones shown in Figure 3 regarding their cytotoxicity, genotoxicity, oxidative damage and ability to induce apoptosis in multiple tumoral cell lines (HepG2, NRK, Caco-2, A549).

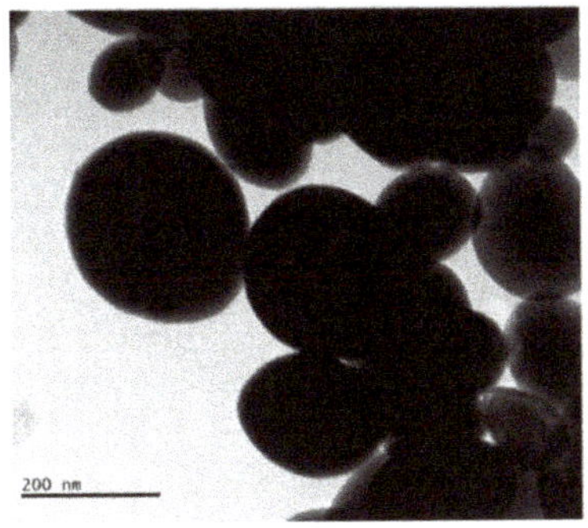

Figure 3. Transmission electronic microscopic capture of bismuth oxide nanoparticles with average diameter ranging from 150 nm to 200 nm. Picture is reprinted with permission from Abudayyak et al. [57].

Authors proved that bismuth oxide nanoparticles differently interacted with different cell lines inducing death through apoptosis in HepG2 and NRK-52E cells and through necrosis in A549 and Caco-2 cells. Among all morphologies, spherical nanoparticles are the most investigated but several studies [58–60] have proved that rod-like particles have a higher cellular uptake and transport across intestinal cells. As reported by Truong et al. [61], morphology is a key point to the rational design of biologically active species with cylindrical particles that are the most suitable for tumor accumulation [62]. Among bismuth species, spherical and sponge-like [17] shapes are the most common morphologies but $BiONO_3$ [26] could be produced as road-like nanosized particles. Even if this material has been used only as precursors in inorganic synthesis [63] and for the realization of biosensors [26], it could represent an interesting material to improve the bismuth oxide material cellular uptaking.

Ahamed et al. [64] evaluated the effect of Bi_2O_3 accordingly to the scheme summarized in Figure 4 by using the MCF-7 cell line.

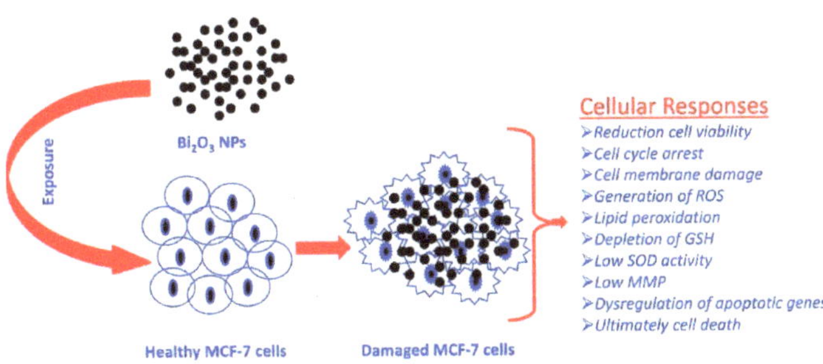

Figure 4. Summary of the process of oxidative stress induced by Bi_2O_3 in MCF-7. Picture is reprinted with permission from Ahamed et al. [64].

The authors observed that the bismuth oxide nanoparticles induced apoptotic response in MCF-7cells and suggested this occurs by undermining the regulation of Bcl-2, Bax and caspase-3 genes. Curiously, the authors observed that with the addition of the external antioxidant N-acetyl-cysteine, the bismuth cytotoxicity was almost inhibited. This suggests that the toxicity of bismuth could be tuned by tailoring the composition of the administered formulation.

Considering the study reported by Thomas et al. [65], bismuth-based drugs could show toxicity to human cells even if they are considered sufficiently safe with careful use. The authors individuated the bismuth methylated species as the main cause responsible for the biological damages induced by bismuth administration due to the increased bioavailability.

Genotoxicity of Bi_2O_3 was also investigated by Liman [66] showing an unneglectable effect on root cells of *Allium cepa*. Even in combination with Portland cement [67] or other minerals [68], bismuth oxide shows a proved citoxicity and antimicrobial effects during in vivo tests.

Li et al. [69] studied the action mechanism of bismuth-based drugs for treating the Helicobacter Pylori infection by using pharmacology and metalloproteomics approaches. The authors described the efficacy of bismuth-based drugs as a consequence of bismuth ability to interrupt several biological pathways by perturbing the activity of key enzymes as shown in Figure 5. The authors suggested that bismuth materials showed the ability to inhibit metallo-β-lactamase by displacing Zn(II) cofactor and proved useful in the treatment of *Helicobacter pylori* infection.

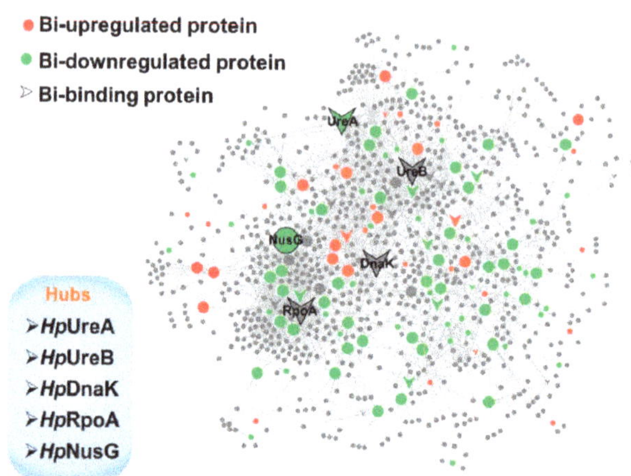

Figure 5. Schematic network depicting bismuth interaction with proteins in *H. pylori*. Proteins are colored and shaped according to their different properties in the network. Adapted with permission from Li et al. [69]. Copyright 2019 American Chemical Society.

Liu et al. [70] explained the toxicity of bismuth-related materials as a consequence of the induction of autophagy in kidney cells. In a previous publication, Liu and co-workers [71] evaluated the bismuth oxide nanoparticles cytotoxicity in human embryonic kidney 293 cells. The authors clarified that autophagy bismuth nanoparticles induced cytotoxicity in kidney cells due to the bismuth ion release from nanoparticles. These bismuth ions altered epigenetically the cells through downregulation of DNA methylation of several gene families. The authors also reported that bismuth nanoparticles were uptaken by cells through non-clathrin-regulated endocytosis with an initial internalization into endosomal compartments with further conversion to lysosomes. This step is the key point of the overall bismuth oxides activity because only after the cell internalization the nanoparticles could exploit their features [72]. To facilitate the uptake, several approaches have been proposed but they are counterbalanced by the reduction of bismuth activity due to the stabilization of nanoparticles. For example, Staedler et al. [73] evaluated the cellular uptake and biocompatibility of bismuth ferrite nanoparticles by using A549, NCI-H520 and THP-1 cell lines. The authors showed consolidating results regarding the depletion of cytotoxicity, haemolytic response and biocompatibility enhancement when nanoparticles are coated with a surfactant like poly(ethylenglycol). Tsang et al. [74] tracked the bismuth in *H. pylori* proving that in some cases bismuth materials could enter the cell through the same metallo-protein complexes accountable for the iron uptake.

Moving on from neat bismuth oxides, another interesting bioactive bismuth species is represented by bismuth oxohalides. Gao et al. [75] reported an in vitro study on the cytotoxicity of BiOCl nanosheets in human HaCaT keratinocytes. The authors reported negligible BiOCl cytotoxicity for concentrations lower than 0.5 μg/mL but the appreciable effect on cancerous cells for concentrations ranging from 5 μg/mL of up to 100 μg/mL. The authors related the cytotoxicity of BiOCl with changes in cell morphology and impairment of intracellular organules. Furthermore, BiOCl induced apoptosis through oxidative stress and eventually cells cycle arrest in G0/G1 phase.

Several proves have been reported on the combination of BiOI photocatalytic activity and antimicrobial effect as described by Jamil et al. [76] and outlined in Figure 6 for the inhibition of *Escherichia coli*.

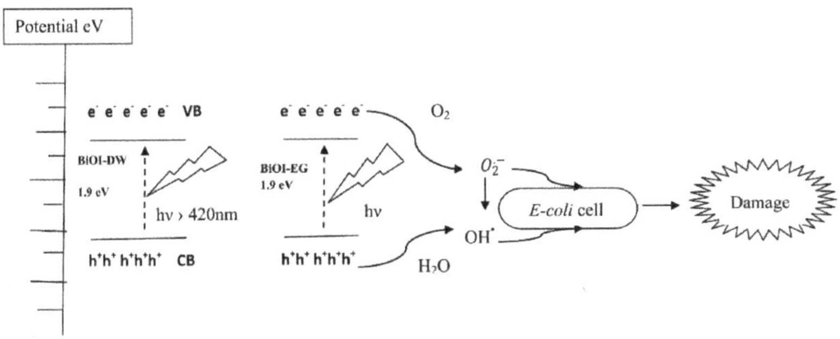

Figure 6. Schematic illustration of energy bands, electron–hole separation and damage mechanism for *E. coli*. Adapted with permission from Jamil et al. [76].

Authors developed solvothermal template synthesis for the production of flower-like structure as those described for BiOX (X = Cl, Br, I) [15] with a high surface area of up to 410 m^2/g. Using a catalyst loading of 0.75 g/L in a watery medium, the authors reported a complete photocatalytic inactivation of *E. coli* strain in a concentration of up to 10^5 CFU/mL.

Similarly, Hsu et al. [77] synthesized gold-doped BiOI nanocomposites through a simple room temperature procedure in an aqueous medium. Gold doped BiOI nanosheets are particularly attractive due to the oxygen vacancies generated in the BiOI lattice structure that increase the oxidation activity. Additionally, the presence of gold nanoparticles enhanced the overall oxidative activity leading to a compound that shows antimicrobial efficacy against *E. coli*, *Klebsiella pneumoniae*, *Salmonella enteritidis*, *Bacillus subtilis* and methicillin-resistant *Staphilococcus aureus*, a methicillin-resistant bacterium. The gold nanoparticles conjugation reduced by up to two-thousand times the minimal inhibitory concentration compared with neat BiOI nanoparticles. The authors also established the pathways of doped BiOI antimicrobial activity that was due to a combined effect of the disruption of the bacterial membrane and the generation of reactive oxygen species. Furthermore, in vivo rabbit model test showed a relevant therapeutic benefit for alleviating corneal *S.aureus* infection without causing inflammatory tissue responses.

Yang et al. [78] reported a study on phototherapy ablation of rabbit orthotopic tumors by using non-stoichiometric $BiPO_{4-x}$ nanoparticles. Through near-infrared light absorption, these oxygen defective structures *promoted* hyperthermia together with the formation of reactive oxygen species. Consequently, they were tested for photothermal/photodynamic therapy in vivo using rabbit as the macroanimal model.

Direct effects of bismuth oxide and related materials on cell viability are not the only appreciable strategies that could base on these materials. Bismuth oxides could be used as effective radiosensitizers species. A radiosensitizer is a chemical that increases the radiation effect on cell viability. These chemotherapy agents are used during radiotherapy in combination with harmful radiation to damage the DNA of cells. As reported by Lawrence [79], radiosensitizing represents the greatest step forward in anticancer treatment and nanoparticle species are one of the most interesting materials for such aim [80]. In 2016, Stewart and co-workers [81] reported the first case of study of bismuth oxide nanoparticles as efficient radiosensitizers on highly radioresistant 9L gliosarcoma cell line. The authors exposed 9 L cells to a bismuth oxide nanoparticle concentration of up to 50 μg/mL achieving a sensitization enhancement of up to 1.5 and 1.3 by using an energy of 125 kV and 10 MV, respectively. Similarly, Liu et al. [82] combined radiotherapy and chemotherapy treatments by administration of mesoporous bismuth litchi-shaped $Na_{0.2}Bi_{0.8}O_{0.35}F_{1.91}$ as both radiosensitizer and as a nanovehicle for loading and slow-releasing doxorubicin. This bismuth oxide material combined with radiation and doxorubicin showed a remarkable synergistic ability for tumor elimination ability. Farahani et al. [83] combined the bismuth nanoparticles with polymer gel dosimetry technique testing their effect in kilovolt and Megavolt radiation therapy proving the strong energy dependence of dose enhancement.

3.2. Bismuth Based Nanomaterials as Additives for the Production of Biomaterials

Bismuth oxides and related materials are quite interesting for all the applications where a high radiopacity together with a good value of biocompatibility is required. Radiopacity is simply defined according to the following equation [84]:

$$I_{(x)} = I_0^{-\rho\mu_{(v)}x} \tag{1}$$

where $I_{(x)}$ is the intensity of the attenuated radiation, I_0 is the original radiation intensity, ρ is the mass density of the material, $\mu_{(v)}$ is the attenuation coefficient for a fixed radiation frequency and x is the length of the travelled path through the material.

For biological applications, radiopacity is measured by using the Hounsfield scale [85] according to the following equation

$$\text{Radiopacity} = 10^3 \cdot \frac{\mu_{(v)} - \mu_w}{\mu_w - \mu_a} \tag{2}$$

where μ_w is the attenuation coefficient of water and μ_a is the attenuation coefficient of air.

Bismuth based materials have raised great interest in the production of orthodontic cement due to a combination of biocompatibility, radiopacity and antimicrobial effects [86,87].

da Silveira Bueno et al. [88] studied the composition of Bi_2O_3 containing Portland cement aiming to obtain an adequate radiopacity for endodontic use. The authors mixed Portland cement with the bismuth oxide with concentrations ranging from 5 wt.% of up to 30 wt.% comparing the results with aluminum foils. Materials reached a radiopacity ranging from 0 to 255 on the Hounsfield scale and a value compatible with shielding applications by using a bismuth concentration of up to 15 wt.%.

Similarly. Chen et al. [89] combined Portland cement not only with Bi_2O_3 but also with zirconia. The authors prepared the hybrid bismuth oxide/zirconia compound through a solid-state synthesis at 700 °C for 12 h and mixed it with cement and calcium sulphate. The results showed that the bismuth oxide/zirconia containing cement exploited a greater radiopacity together with the same cell viability of zirconia free one.

Coutinho-Filho et al. [90] evaluated through a histological assessment the subcutaneous connective tissue reactions and the radiopacity of Portland cement mixed with bismuth oxide. The authors reported complete biocompatibility in vivo after 7 and 60 days (no tissue damage observed).

Similarly, several authors reported analogous results for dental repairing applications performed by using Portland and bismuth oxide composites proofing their reliability [91,92].

Furthermore, bismuth could be used for a tissue engineering application as reported by Pazarçeviren et al. [93]. The authors doped a composite made of 45S5 nanobioactive bioglass and graphene oxide with bismuth nanoparticles through a sol–gel methodology. By adding bismuth, authors increased both the composite density and the diametral tensile strength of up to 2.5% retaining cell viability. Additionally, bismuth oxides and related materials could be dispersed into a polymeric matrix to mitigate the effect of harmful radiations during the diagnostic procedures [94,95].

3.3. *Bismuth Based Nanomaterials as Diagnostic Agents*

Bismuth oxides and related materials are also used as contrast agents due to their radiopacity. Bi et al. [96] used poly(ethylenglycol) modified bismuth nanoparticles for applications as multifunctional probes during X-ray computed tomography (CT) and fluorescence imaging. The authors tested the in vivo circulation time and specific accumulation behavior in the liver and intestines by using a CT scan as shown in Figure 7.

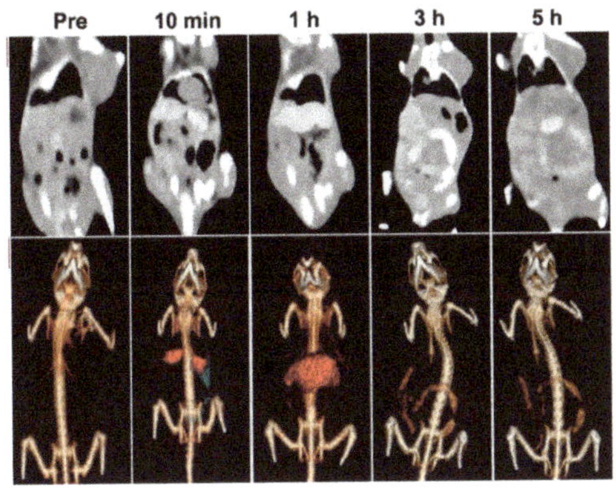

Figure 7. CT images and renderings of CT images of rat after the administration of bismuth modified nanoparticles after different times. Adapted with permission from Bi et al. [96]. Copyright 2018 American Chemical Society.

Results showed the possible applications of these formulations for target imaging and tracing of the specific areas where bismuth was preferentially accumulated.

Similarly, Swy et al. [97] produced poly(lactic-co-glycolic acid) encapsulated bismuth nanoparticles with an average diameter of less than 40 nm using them as fluorescent probes. The authors achieved a degradation of bismuth-based probes of up to 90% in the acidic and lysosomal-like environment after 24 h while they remained in cytosolic and extracellular-like fluid media.

Several studies have proved the reliability of bismuth oxide as a CT contrast agent with similar or better performances compared with other oxides [98]. Brown et al. [99] developed an ultra-high payload metallic bismuth nanoparticle used as X-ray contrast agents. The authors showed that metallic bismuth nanoparticles will oxidatively decompose to biocompatible Bi(III) based species that are renal excreted after the CT analysis. Dadashi and co-workers [100] combined bismuth nanoparticles together with gold species producing aggregates of up to 40 nm in diameter demonstrating a higher X-ray attenuation in comparison with commercial iodine-based molecules.

Hu et al. [101] synthesized a nanostructured $(BiO)_2CO_3$ rod-like material through a solvothermal route and used it as a renal clearable CT contrast agent as shown in Figure 8.

The authors efficiently used the bismuth subcarbonate as a high-resolution CT contrast agent proving that its high aspect ratio actively promoted take-up and retention in the rat tumors tested. The authors also reported the disassembling of the bismuth rods in the acidic microenvironment of tumors enhancing the renal clearance.

Naha et al. [102] reported the production of dextran-coated bismuth/iron oxide nanostructures for magnetic resonance (MR) applications. Results showed a decrement in T2-weighted MR contrast with increasing bismuth content in liver cells. The authors did not observe any cytotoxicity on Hep G2 and BJ5ta cell lines after 24 h incubation with the nanohybrids. Furthermore, the authors ran an in vivo test using mice observing a 2 h circulation time in heart and blood vessels of the bismuth contrast agent. Additionally, this bismuth-based contrast agent was rapidly excreted with urine.

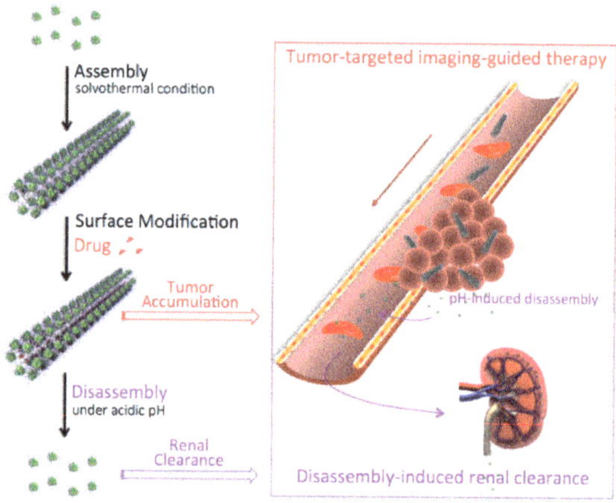

Figure 8. Production and biological pathway of bismuth subcarbonates rods as reported by [101]. Copyright 2018 American Chemical Society.

Rivera et al. [103] encapsulated BiOCl into carbon nanostructures and tested it as an agent for X-ray imaging. The authors achieved a high contrast by using a low bismuth loading on nanocarbon (up to 2.7 wt.%) without compromising cell viability. Data enlightened a magnification of up to 500 times of CT resolution compared with traditional iodine-based agents.

BiOCl could be also used as support for the immobilization of aptameric tailored gold nanoparticles as reported by Hsu et al. [104]. This hybrid material showed high peroxidase-like activity and was used for the conversion of Amplex Red proteinic complex to resorufin. According to the authors, this was a very remarkable achievement that proved the robustness of bismuth bioconjugate in proteomic applications.

3.4. *Bismuth Based Nanomaterials as Active Agents in Theragnostic Platforms*

The combination of diagnostic procedures together with a therapeutic protocol is defined as theragnostic and represents the last frontier in advanced treatments [105]. Nanoscale theragnostic is a fast-growing branch of medicinal chemistry for simultaneously monitoring drug release and its distribution, and to evaluate the real-time therapeutic efficacy through a single nanoscale product for both treatment and diagnosis. As reported in the previous sections, bismuth materials are good and efficient contrast agents but could also be exploited for targeted cytotoxicity in vivo. The simultaneous effects herein mentioned led to the development of theragnostic platforms based on bismuth oxides and related materials.

Li et al. [106] developed a bovine serum albumin modified bismuth oxides nanoraspberries for multimodal imaging and chemo-photothermal combination therapy as summarized in Figure 9.

The authors synthesized the nanoparticles through a watery reduction by using sodium borhydride under pressure at 150 °C for 3 h. The synthesized material showed a surface area of up to 53 m^2/g and a DOX drug loading of up to 69 wt.% with release occurring upon pH variations. The authors reported the bismuth-based theragnostic agent's ability to efficiently convert near-infrared light to thermal energy for photothermal ablation of cancer cells. The toxicity studies proved the high biocompatibility without any appreciable toxicity to the mice tested. Additionally, the high radiopacity of bismuth raspberries allows the use of this formulation also during CT analysis. Lu et al. using a similar approach combined the radiopacity of bismuth nanoparticles with photothermal therapy. The authors were able to reach up to 70 °C after 4 min of infrared irradiation showing an enhancement in both CT imaging and in vitro suppression of glioma growth. Xuan et al. [107] prepared bismuth nanoparticles embedded into a nanohydrogel by ultraviolet light-mediated synthesis. The produced materials were combined with DOX and used simultaneously as a contrast agent, as a nanocarrier for drugs and for inducing cell death by thermal ablation. Analogously, Yang et al. [108] produced a bismuth-based CT contrast agent used in photothermal therapy and in ultrasound imaging. They used also several tailored approaches aimed to enhance the theragnostic effects of bismuth preparations. Yu et al. [109] described a thiol capping of bismuth nanoparticles that prevents the unwanted release of bismuth in the organism.

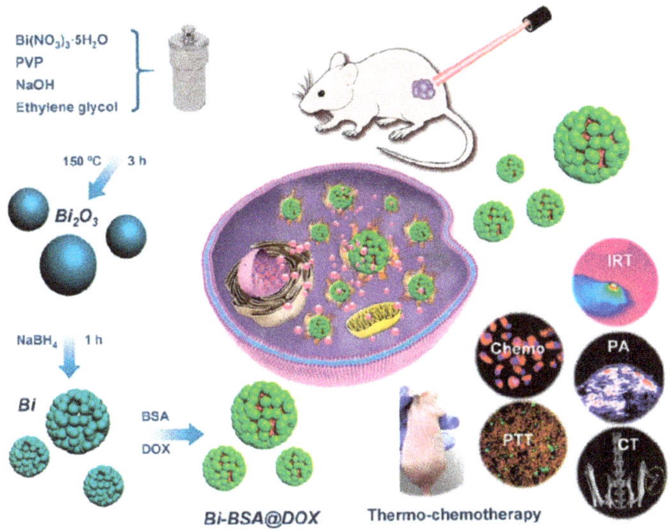

Figure 9. Production and biological action of bismuth oxide nanoraspberries species as reported by Li et al. [106]. Copyright 2018 American Chemical Society.

Bismuth oxides and related materials could be also combined with other species. Detappe et al. [110] produced a hybrid material by using ultrasmall silica-based bismuth and gadolinium nanoparticles for dual magnetic resonance and CT imaging while Badrigilan et al. [111] conjugated Bi_2O_3 with iron oxides to improve the photothermal behaviour leaving untouched the high bismuth radiopacity.

4. Conclusions

Bismuth oxides and related materials show a unique set of features that harvest a relevant interest for plenty of biological applications. Ranging from the production of active drugs to diagnostic agents, bismuth could play a major role in the extension of these productions enhancing the state-of-the-art limit reaching new goals.

While there are plenty of more effective and specific drugs, the combination of radiopacity and tunability of bismuth is quite a unique combination of properties. This represents the starting point for the development of theragnostic platforms that could represent a real game-changing event in the field of advanced medicine. Theragnostic is the field where bismuth could exploit its full potential. Nonetheless, the main unaddressed challenge in the biological application of bismuth is represented by the preparation of an only bismuth-based theragnostic platform where bismuth is simultaneously the contrast agent and the bio-active specie. On this very same topic, a highly speculative but realistic approach could be represented by the realization of a bismuth multilayered particles where the core of bismuth oxide is covered by bismuth oxynitrate tailored with specific biological markers. This hypothesized specie could be the first self-standing mono-element theragnostic preparation where bismuth oxide provides the radiopacity and surface tailoring together with intrinsic defects could provide specific cytotoxicity and drug delivery system.

We strongly believe that the research of new bismuth oxide-based formulations and nanoarchitectures will lead to major breakthroughs with a huge positive impact on humankind's welfare.

Author Contributions: Conceptualization, M.B., P.J. and A.T; writing—original draft preparation, M.B., P.J. and A.T.; writing—review and editing, A.T., P.J. and M.B.; supervision, A.T. All authors have read and agreed to the published version of the manuscript.

Funding: This research received no external funding.

Conflicts of Interest: The authors declare no conflict of interest.

References

1. Girish, G.; Finlay, K.; Fessell, D.; Pai, D.; Dong, Q.; Jamadar, D. Imaging review of skeletal tumors of the pelvis malignant tumors and tumor mimics. *Sci. World J.* **2012**, *2012*, 240281. [CrossRef] [PubMed]
2. Boone, C.W.; Kelloff, G.J.; Malone, W.E. Identification of candidate cancer chemopreventive agents and their evaluation in animal models and human clinical trials: A review. *Cancer Res.* **1990**, *50*, 2–9. [PubMed]
3. Tannock, I. Cell kinetics and chemotherapy: A critical review. *Cancer Treat. Rep.* **1978**, *62*, 1117. [PubMed]
4. Zaman, S.B.; Hussain, M.A.; Nye, R.; Mehta, V.; Mamun, K.T.; Hossain, N. A review on antibiotic resistance: Alarm bells are ringing. *Cureus* **2017**, *9*, e1403. [CrossRef] [PubMed]
5. Shahbazi, M.-A.; Faghfouri, L.; Ferreira, M.P.; Figueiredo, P.; Maleki, H.; Sefat, F.; Hirvonen, J.; Santos, H.A. The versatile biomedical applications of bismuth-based nanoparticles and composites: Therapeutic, diagnostic, biosensing, and regenerative properties. *Chem. Soc. Rev.* **2020**, *49*, 1253–1321. [CrossRef] [PubMed]
6. Monseu, G.; Struelens, M.; Roland, M. Bismuth encephalopathy. *Acta Neurol. Belg.* **1976**, *76*, 301–308. [PubMed]
7. Buge, A.; Supino-Viterbo, V.; Rancurel, G.; Pontes, C. Epileptic phenomena in bismuth toxic encephalopathy. *J. Neurol. Neurosurg. Psychiatry* **1981**, *44*, 62–67. [CrossRef]
8. DuPont, H.L.; Ericsson, C.D.; Johnson, P.C.; Bitsura, J.A.M.; DuPont, M.W.; de la Cabada, F.J. Prevention of travelers' diarrhea by the tablet formulation of bismuth subsalicylate. *JAMA* **1987**, *257*, 1347–1350. [CrossRef]
9. Bierer, D.W. Bismuth subsalicylate: History, chemistry, and safety. *Rev. Infect. Dis.* **1990**, *12*, S3–S8. [CrossRef]
10. Wagstaff, A.J.; Benfield, P.; Monk, J.P. Colloidal bismuth subcitrate. *Drugs* **1988**, *36*, 132–157. [CrossRef]
11. Goodwin, C.; Marshall, B.; Blincow, E.; Wilson, D.; Blackbourn, S.; Phillips, M. Prevention of nitroimidazole resistance in Campylobacter pylori by coadministration of colloidal bismuth subcitrate: Clinical and in vitro studies. *J. Clin. Pathol.* **1988**, *41*, 207–210. [CrossRef] [PubMed]
12. Wieriks, J.; Hespe, W.; Jaitly, K.; Koekkoek, P.; Lavy, U. Pharmacological properties of colloidal bismuth subcitrate (CBS, DE-NOL). *Scand. J. Gastroenterol. Suppl.* **1982**, *80*, 11. [PubMed]
13. Canena, J.; Reis, J.; Pinto, A.; Santos, A.; Leitao, J.; Pinheiro, T.; Quina, M. Distribution of bismuth in the rat after oral dosing with ranitidine bismuth citrate and bismuth subcitrate. *J. Pharm. Pharm.* **1998**, *50*, 279–283. [CrossRef] [PubMed]
14. Barrera-Mota, K.; Bizarro, M.; Castellino, M.; Tagliaferro, A.; Hernández, A.; Rodil, S.E. Spray deposited β-Bi_2O_3 nanostructured films with visible photocatalytic activity for solar water treatment. *Photochem. Photobiol. Sci.* **2015**, *14*, 1110–1119. [CrossRef]
15. Jagdale, P.; Castellino, M.; Marrec, F.; Rodil, S.E.; Tagliaferro, A. Nano sized bismuth oxy chloride by metal organic chemical vapour deposition. *Appl. Surf. Sci.* **2014**, *303*, 250–254. [CrossRef]
16. Meitl, M.A.; Dellinger, T.M.; Braun, P.V. Bismuth–Ceramic Nanocomposites with Unusual Thermal Stability via High-Energy Ball Milling. *Adv. Funct. Mater.* **2003**, *13*, 795–799. [CrossRef]
17. Gadhi, T.A.; Hernández, S.; Castellino, M.; Jagdale, P.; Husak, T.; Hernández-Gordillo, A.; Tagliaferro, A.; Russo, N. Insights on the role of β-Bi_2O_3/$Bi_5O_7NO_3$ heterostructures synthesized by a scalable solid-state method for the sunlight-driven photocatalytic degradation of dyes. *Catal. Today* **2019**, *321–322*, 135–145. [CrossRef]
18. Gadhi, T.A.; Hernández-Gordillo, A.; Bizarro, M.; Jagdale, P.; Tagliaferro, A.; Rodil, S.E. Efficient α/β-Bi_2O_3 composite for the sequential photodegradation of two-dyes mixture. *Ceram. Int.* **2016**, *42*, 13065–13073. [CrossRef]

19. Singh, N.; Singh, K.J.; Singh, K.; Singh, H. Comparative study of lead borate and bismuth lead borate glass systems as gamma-radiation shielding materials. *Nucl. Instrum. Methods Phys. Res. Sect. B* **2004**, *225*, 305–309. [CrossRef]
20. Mallaley, K.; Morris, D. Analysis of the Betterton-Kroll Process: The Removal of Bismuth from Lead Bullion. In *Primary and Secondary Lead Processing*; Elsevier: Amsterdam, The Netherlands, 1989; pp. 253–262.
21. González-Domínguez, J.; Peters, E.; Dreisinger, D. The refining of lead by the Betts process. *J. Appl. Electrochem.* **1991**, *21*, 189–202. [CrossRef]
22. Zhou, Y.; Dong, F.; Jin, S. *Bismuth: Advanced Applications and Defects Characterization*; InTech: London, UK, 2018.
23. Berry, C.R. Electron diffraction from small crystals. *Phys. Rev.* **1952**, *88*, 596. [CrossRef]
24. Zhao, L.; Zhang, X.; Fan, C.; Liang, Z.; Han, P. First-principles study on the structural, electronic and optical properties of BiOX (X = Cl, Br, I) crystals. *Phys. B* **2012**, *407*, 3364–3370. [CrossRef]
25. Greenwood, N.N.; Earnshaw, A. *Chemistry of the Elements*; Elsevier: Amsterdam, The Netherlands, 2012.
26. Liu, G.-Q.; Zhong, H.; Li, X.-R.; Yang, K.; Jia, F.-F.; Cheng, Z.-P.; Zhang, L.-L.; Yin, J.-Z.; Guo, L.-P.; Qian, H.-Y. Research on nonenzymatic electrochemical sensor using HO-BiONO3 nanocomposites for glucose detection. *Sens. Actuators B* **2017**, *242*, 484–491. [CrossRef]
27. Briand, G.G.; Burford, N. Bismuth compounds and preparations with biological or medicinal relevance. *Chem. Rev.* **1999**, *99*, 2601–2658. [CrossRef] [PubMed]
28. Lu, B.; Zhu, Y. Synthesis and photocatalysis performances of bismuth oxynitrate photocatalysts with layered structures. *PCCP* **2014**, *16*, 16509–16514. [CrossRef]
29. Yang, Y.; Liang, H.; Zhu, N.; Zhao, Y.; Guo, C.; Liu, L. New type of $[Bi_6O_6(OH)_3](NO_3)_3{\cdot}1.5H_2O$ sheets photocatalyst with high photocatalytic activity on degradation of phenol. *Chemosphere* **2013**, *93*, 701–707. [CrossRef]
30. Zahariev, A.; Kaloyanov, N.; Girginov, C.; Parvanova, V. Synthesis and thermal decomposition of $[Bi_6O_6(OH)_2](NH_2C_6H_4SO_3)_4$. *Thermochim. Acta* **2012**, *528*, 85–89. [CrossRef]
31. Kodama, H. Synthesis of a new compound, $Bi_5O_7NO_3$, by thermal decomposition. *J. Solid State Chem.* **1994**, *112*, 27–30. [CrossRef]
32. Yu, S.; Zhang, G.; Gao, Y.; Huang, B. Single-crystalline $Bi_5O_7NO_3$ nanofibers: Hydrothermal synthesis, characterization, growth mechanism, and photocatalytic properties. *J. Colloid Interface Sci.* **2011**, *354*, 322–330. [CrossRef]
33. Levin, E.M.; Roth, R.S. Polymorphism of bismuth sesquioxide. I. Pure Bi_2O_3. *J. Res. Natl. Bur. Stand. Sect. A Phys. Chem.* **1964**, *68*, 189. [CrossRef]
34. Lei, B.; Cui, W.; Sheng, J.; Wang, H.; Chen, P.; Li, J.; Sun, Y.; Dong, F. Synergistic effects of crystal structure and oxygen vacancy on Bi_2O_3 polymorphs: Intermediates activation, photocatalytic reaction efficiency, and conversion pathway. *Sci. Bull.* **2020**, *65*, 467–476. [CrossRef]
35. Bruton, T.; Brice, J.; Hill, O.; Whiffin, P. The flux growth of some γ-Bi_2O_3 crystals by the top seeded technique. *J. Cryst. Growth* **1974**, *23*, 21–24. [CrossRef]
36. Gualtieri, A.; Immovilli, S.; Prudenziati, M. Powder X-ray diffraction data for the new polymorphic compound ω-Bi_2O_3. *Powder Diffr.* **1997**, *12*, 90–92. [CrossRef]
37. Cornei, N.; Tancret, N.; Abraham, F.; Mentré, O. New ε-Bi_2O_3 metastable polymorph. *Inorg. Chem.* **2006**, *45*, 4886–4888. [CrossRef]
38. Kumada, N.; Takahashi, N.; Kinomura, N.; Sleight, A. Preparation and crystal structure of a new lithium bismuth oxide: $LiBiO_3$. *J. Solid State Chem.* **1996**, *126*, 121–126. [CrossRef]
39. Zhang, T.; Ding, Y.; Tang, H. Generation of singlet oxygen over Bi (V)/Bi (III) composite and its use for oxidative degradation of organic pollutants. *Chem. Eng. J.* **2015**, *264*, 681–689. [CrossRef]
40. Kiran, V.S.; Sumathi, S. Comparison of catalytic activity of bismuth substituted cobalt ferrite nanoparticles synthesized by combustion and co-precipitation method. *J. Magn. Magn. Mater.* **2017**, *421*, 113–119. [CrossRef]
41. Petsom, K.; Kopwitthaya, A.; Horphathum, M.; Kaewkhao, J.; Sangwaranatee, N. The effect of additive chemicals on synthesis of bismuth nanoparticles. *Mater. Today Proc.* **2018**, *5*, 14057–14062. [CrossRef]
42. Tramontina, V.A.; Machado, M.A.N.; Filho, G.d.R.N.; Kim, S.H.; Vizzioli, M.R.; Toledo, S. Effect of bismuth subgallate (local hemostatic agent) on wound healing in rats. Histological and histometric findings. *Braz. Dent. J.* **2002**, *13*, 11–16.

43. Pamphlett, R.; Stoltenberg, M.; Rungby, J.; Danscher, G. Uptake of bismuth in motor neurons of mice after single oral doses of bismuth compounds. *Neurotoxicol. Teratol.* **2000**, *22*, 559–563. [CrossRef]
44. Slikkerveer, A.; de Wolff, F.A. Pharmacokinetics and toxicity of bismuth compounds. *Med. Toxicol. Advers. Drug Exp.* **1989**, *4*, 303–323. [CrossRef] [PubMed]
45. Nwokolo, C.U.; Gavey, C.J.; Smith, J.T.L.; Pounder, R.E. The absorption of bismuth from oral doses of tripotassium dicitrato bismuthate. *Aliment. Pharm. Ther.* **1989**, *3*, 29–39. [CrossRef] [PubMed]
46. Conso, F. Bismuth Sanguin Et Urinaire Apres Traitement Bref Par Differents Sels Insolubles De Bismuth. *Eur. J. Toxicol.* **1975**, *8*, 137–141.
47. Gavey, C.; Szeto, M.L.; Nwokolo, C.; Sercombe, J.; Pounder, R. Bismuth accumulates in the body during treatment with tripotassium dicitrato bismuthate. *Aliment. Pharm. Ther.* **1989**, *3*, 21–28. [CrossRef] [PubMed]
48. Chaleil, D. Augmentation des concentrations sanguines de bismuth par la cystéine chez le rat. *Thérapie* **1979**, *34*, 397–399.
49. Lechat, P.; Majoie, B.; Levillain, R.; Cluzan, R.; Deleau, D. Étude de la toxicité à court terme de l'association sous-nitrate de bismuth et sorbitol. *Therapie* **1964**, *19*, 551–556.
50. Szymanska, J.A.; Mogilnicka, E.M.; Kaszper, B.W. Binding of bismuth in the kidney of the rat. The role of metallothionein-like proteins. *Biochem. Pharmacol.* **1977**, *26*, 257–258. [CrossRef]
51. Szymanska, J.A.; Piotrowski, J.K. Studies to identify the low molecular weight bismuth-binding proteins in rat kidney. *Biochem. Pharm.* **1980**, *29*, 2913–2918. [CrossRef]
52. Żelazowski, A.J.; Piotrowski, J.K. Mercury-binding, copper-zinc proteins from rat kidney. Amino acid composition, molecular weight and metal content. *Biochim. Et Biophys. Acta (Bba)-Protein Struct.* **1980**, *625*, 89–99. [CrossRef]
53. Chaleil, D.; Regnault, J.; Allain, P.; Motta, R.; Raynaud, G. Action d'une flore microbienne méthanogène d'origine humaine sur l'absorption et la fixation du bismuth chez le rat. *Ann. Pharm. Fr.* **1988**, *46*, 133–137.
54. Buge, A. 20 observations d'encéphalopathies aiguës avec myoclonies au cours de traitements oraux par les sels de bismuth. *Ann. Méd. Interne* **1974**, *125*, 877–888.
55. Stephens, L.J.; Munuganti, S.; Duffin, R.N.; Werrett, M.V.; Andrews, P.C. Is Bismuth Really the "Green" Metal? Exploring the Antimicrobial Activity and Cytotoxicity of Organobismuth Thiolate Complexes. *Inorg. Chem.* **2020**, *59*, 3494–3508. [CrossRef] [PubMed]
56. Luo, Y.; Wang, C.; Qiao, Y.; Hossain, M.; Ma, L.; Su, M. In vitro cytotoxicity of surface modified bismuth nanoparticles. *J. Mater. Sci. Mater. Med.* **2012**, *23*, 2563–2573. [CrossRef] [PubMed]
57. Abudayyak, M.; Öztaş, E.; Arici, M.; Özhan, G. Investigation of the toxicity of bismuth oxide nanoparticles in various cell lines. *Chemosphere* **2017**, *169*, 117–123. [CrossRef] [PubMed]
58. Banerjee, A.; Qi, J.; Gogoi, R.; Wong, J.; Mitragotri, S. Role of nanoparticle size, shape and surface chemistry in oral drug delivery. *J. Control. Release* **2016**, *238*, 176–185. [CrossRef] [PubMed]
59. Mitragotri, S. In drug delivery, shape does matter. *Pharm. Res.* **2009**, *26*, 232–234. [CrossRef] [PubMed]
60. Christian, D.A.; Cai, S.; Garbuzenko, O.B.; Harada, T.; Zajac, A.L.; Minko, T.; Discher, D.E. Flexible filaments for in vivo imaging and delivery: Persistent circulation of filomicelles opens the dosage window for sustained tumor shrinkage. *Mol. Pharm.* **2009**, *6*, 1343–1352. [CrossRef]
61. Truong, N.P.; Whittaker, M.R.; Mak, C.W.; Davis, T.P. The importance of nanoparticle shape in cancer drug delivery. *Expert Opin. Drug Deliv.* **2015**, *12*, 129–142. [CrossRef]
62. Dickerson, E.B.; Dreaden, E.C.; Huang, X.; El-Sayed, I.H.; Chu, H.; Pushpanketh, S.; McDonald, J.F.; El-Sayed, M.A. Gold nanorod assisted near-infrared plasmonic photothermal therapy (PPTT) of squamous cell carcinoma in mice. *Cancer Lett.* **2008**, *269*, 57–66. [CrossRef]
63. Zhou, C.; Cao, J.; Lin, H.; Xu, B.; Huang, B.; Chen, S. Controllable synthesis and photocatalytic activity of Ag/BiOI based on the morphology effect of BiOI substrate. *Surf. Coat. Technol.* **2015**, *272*, 213–220. [CrossRef]
64. Ahamed, M.; Akhtar, M.J.; Khan, M.A.M.; Alrokayan, S.A.; Alhadlaq, H.A. Oxidative stress mediated cytotoxicity and apoptosis response of bismuth oxide (Bi_2O_3) nanoparticles in human breast cancer (MCF-7) cells. *Chemosphere* **2019**, *216*, 823–831. [CrossRef] [PubMed]
65. Thomas, F.; Bialek, B.; Hensel, R. Medical use of bismuth: The two sides of the coin. *J. Clin. Toxicol.* **2012**, *3*, 004.
66. Liman, R. Genotoxic effects of Bismuth (III) oxide nanoparticles by Allium and Comet assay. *Chemosphere* **2013**, *93*, 269–273. [CrossRef] [PubMed]

67. Zeferino, E.; Bueno, C.S.; Oyama, L.; Ribeiro, D. Ex vivo assessment of genotoxicity and cytotoxicity in murine fibroblasts exposed to white MTA or white Portland cement with 15% bismuth oxide. *Int. Endod. J.* **2010**, *43*, 843–848. [CrossRef] [PubMed]
68. Camilleri, J.; Montesin, F.E.; Papaioannou, S.; McDonald, F.; Pitt Ford, T.R. Biocompatibility of two commercial forms of mineral trioxide aggregate. *Int. Endod. J.* **2004**, *37*, 699–704. [CrossRef]
69. Li, H.; Wang, R.; Sun, H. Systems Approaches for Unveiling the Mechanism of Action of Bismuth Drugs: New Medicinal Applications beyond Helicobacter Pylori Infection. *Acc. Chem. Res.* **2019**, *52*, 216–227. [CrossRef]
70. Liu, Y.; Shen, C.; Zhang, X.; Yu, H.; Wang, F.; Wang, Y.; Zhang, L.W. Exposure and nephrotoxicity concern of bismuth with the occurrence of autophagy. *Toxicol. Ind. Health* **2018**, *34*, 188–199. [CrossRef]
71. Liu, Y.; Zhuang, J.; Zhang, X.; Yue, C.; Zhu, N.; Yang, L.; Wang, Y.; Chen, T.; Wang, Y.; Zhang, L.W. Autophagy associated cytotoxicity and cellular uptake mechanisms of bismuth nanoparticles in human kidney cells. *Toxicol. Lett.* **2017**, *275*, 39–48. [CrossRef]
72. Reus, T.L.; Machado, T.N.; Bezerra, A.G.; Marcon, B.H.; Paschoal, A.C.C.; Kuligovski, C.; de Aguiar, A.M.; Dallagiovanna, B. Dose-dependent cytotoxicity of bismuth nanoparticles produced by LASiS in a reference mammalian cell line BALB/c 3T3. *Toxicol. Vitr.* **2018**, *53*, 99–106. [CrossRef]
73. Staedler, D.; Passemard, S.; Magouroux, T.; Rogov, A.; Maguire, C.M.; Mohamed, B.M.; Schwung, S.; Rytz, D.; Jüstel, T.; Hwu, S.; et al. Cellular uptake and biocompatibility of bismuth ferrite harmonic advanced nanoparticles. *Nanomed. Nanotechnol. Biol. Med.* **2015**, *11*, 815–824. [CrossRef]
74. Tsang, C.-N.; Ho, K.-S.; Sun, H.; Chan, W.-T. Tracking Bismuth Antiulcer Drug Uptake in Single Helicobacter pylori Cells. *J. Am. Chem. Soc.* **2011**, *133*, 7355–7357. [CrossRef] [PubMed]
75. Gao, X.; Zhang, X.; Wang, Y.; Wang, Y.; Peng, S.; Fan, C. An in vitro study on the cytotoxicity of bismuth oxychloride nanosheets in human HaCaT keratinocytes. *Food Chem. Toxicol.* **2015**, *80*, 52–61. [CrossRef] [PubMed]
76. Jamil, T.S.; Mansor, E.S.; Azab El-Liethy, M. Photocatalytic inactivation of E. coli using nano-size bismuth oxyiodide photocatalysts under visible light. *J. Environ. Chem. Eng.* **2015**, *3*, 2463–2471. [CrossRef]
77. Hsu, C.-L.; Li, Y.-J.; Jian, H.-J.; Harroun, S.G.; Wei, S.-C.; Ravindranath, R.; Lai, J.-Y.; Huang, C.-C.; Chang, H.-T. Green synthesis of catalytic gold/bismuth oxyiodide nanocomposites with oxygen vacancies for treatment of bacterial infections. *Nanoscale* **2018**, *10*, 11808–11819. [CrossRef]
78. Yang, C.; Huang, W.; Gao, Y.; Liu, Z.; An, N.; Mu, W.; Pan, Q.; Yang, B.; Guo, C.; Han, X. Phototherapy ablation of rabbit orthotopic tumors by non-stoichiometric BiPO4−x nanoparticles. *Chem. Eng. J.* **2020**, *386*, 123961. [CrossRef]
79. Lawrence, T.S.; Blackstock, A.W.; McGinn, C. The mechanism of action of radiosensitization of conventional chemotherapeutic agents. *Semin. Radiat. Oncol.* **2003**, *13*, 13–21. [CrossRef]
80. Brun, E.; Sicard-Roselli, C. Actual questions raised by nanoparticle radiosensitization. *Radiat. Phys. Chem.* **2016**, *128*, 134–142. [CrossRef]
81. Stewart, C.; Konstantinov, K.; McKinnon, S.; Guatelli, S.; Lerch, M.; Rosenfeld, A.; Tehei, M.; Corde, S. First proof of bismuth oxide nanoparticles as efficient radiosensitisers on highly radioresistant cancer cells. *Phys. Med.* **2016**, *32*, 1444–1452. [CrossRef]
82. Liu, J.; Deng, Y.; Qin, X.; Li, B.; Zhang, J.; Xu, Y.; Ouyang, R.; Li, Y.; Miao, Y.; Sun, Y. Ultrafast Synthesizing Bismuth Mesoporous Nanolitchi Radiosensitizer Loading High Dose DOX for CT-Guided Enhanced Chemoradiotherapy. *Acs Appl. Mater. Interfaces* **2019**, *11*, 42932–42942. [CrossRef]
83. Farahani, S.; Riyahi Alam, N.; Haghgoo, S.; Shirazi, A.; Geraily, G.; Gorji, E.; Kavousi, N. The effect of bismuth nanoparticles in kilovoltage and megavoltage radiation therapy using magnetic resonance imaging polymer gel dosimetry. *Radiat. Phys. Chem.* **2020**, *170*, 108573. [CrossRef]
84. McNaught, A.D.; Wilkinson, A. *Attenuation Coefficient*; Blackwell Science Oxford: Oxford, UK, 1997; Volume 1669.
85. DenOtter, T.D.; Schubert, J. Hounsfield Unit. In *StatPearls [Internet]*; StatPearls Publishing: Treasure Island, FL, USA, 2019.
86. Deb, S.; Abdulghani, S.; Behiri, J. Radiopacity in bone cements using an organo-bismuth compound. *Biomaterials* **2002**, *23*, 3387–3393. [CrossRef]
87. Chen, F.; Liu, C.; Mao, Y. Bismuth-doped injectable calcium phosphate cement with improved radiopacity and potent antimicrobial activity for root canal filling. *Acta Biomater.* **2010**, *6*, 3199–3207. [CrossRef] [PubMed]

88. Da Silveira Bueno, C.E.; Zeferino, E.G.; Manhães, L.R.C., Jr.; Rocha, D.G.P.; Cunha, R.S.; De Martin, A.S. Study of the bismuth oxide concentration required to provide Portland cement with adequate radiopacity for endodontic use. *Oral Surg. Oral Med. Oral Pathol. Oral Radiol. Endod.* **2009**, *107*, e65–e69. [CrossRef] [PubMed]
89. Chen, X.; Song, J.; Chen, X.; Yang, H. X-ray-activated nanosystems for theranostic applications. *Chem. Soc. Rev.* **2019**, *48*, 3073–3101. [CrossRef]
90. Coutinho-Filho, T.; De-Deus, G.; Klein, L.; Manera, G.; Peixoto, C.; Gurgel-Filho, E.D. Radiopacity and histological assessment of Portland cement plus bismuth oxide. *Oral Surg. Oral Med. Oral Pathol. Oral Radiol. Endod.* **2008**, *106*, e69–e77. [CrossRef]
91. Hwang, Y.-C.; Lee, S.-H.; Hwang, I.-N.; Kang, I.-C.; Kim, M.-S.; Kim, S.-H.; Son, H.-H.; Oh, W.-M. Chemical composition, radiopacity, and biocompatibility of Portland cement with bismuth oxide. *Oral Surg. Oral Med. Oral Pathol. Oral Radiol. Endod.* **2009**, *107*, e96–e102. [CrossRef]
92. Kim, E.-C.; Lee, B.-C.; Chang, H.-S.; Lee, W.; Hong, C.-U.; Min, K.-S. Evaluation of the radiopacity and cytotoxicity of Portland cements containing bismuth oxide. *Oral Surg. Oral Med. Oral Pathol. Oral Radiol. Endod.* **2008**, *105*, e54–e57. [CrossRef]
93. Pazarçeviren, A.E.; Tahmasebifar, A.; Tezcaner, A.; Keskin, D.; Evis, Z. Investigation of bismuth doped bioglass/graphene oxide nanocomposites for bone tissue engineering. *Ceram. Int.* **2018**, *44*, 3791–3799. [CrossRef]
94. Jagdale, P.; Rovere, M.; Ronca, R.; Vigneri, C.; Bernardini, F.; Calzetta, G.; Tagliaferro, A. Determination of the X-ray attenuation coefficient of bismuth oxychloride nanoplates in polydimethylsiloxane. *J. Mater. Sci.* **2020**, *55*, 7095–7105. [CrossRef]
95. Mehnati, P.; Arash, M.; Akhlaghi, P. Bismuth-silicon and bismuth-polyurethane composite shields for breast protection in chest computed tomography examinations. *J. Med. Phys.* **2018**, *43*, 61.
96. Bi, H.; He, F.; Dong, Y.; Yang, D.; Dai, Y.; Xu, L.; Lv, R.; Gai, S.; Yang, P.; Lin, J. Bismuth Nanoparticles with "Light" Property Served as a Multifunctional Probe for X-ray Computed Tomography and Fluorescence Imaging. *Chem. Mater.* **2018**, *30*, 3301–3307. [CrossRef]
97. Swy, E.R.; Schwartz-Duval, A.S.; Shuboni, D.D.; Latourette, M.T.; Mallet, C.L.; Parys, M.; Cormode, D.P.; Shapiro, E.M. Dual-modality, fluorescent, PLGA encapsulated bismuth nanoparticles for molecular and cellular fluorescence imaging and computed tomography. *Nanoscale* **2014**, *6*, 13104–13112. [CrossRef] [PubMed]
98. Ghazanfari, A.; Marasini, S.; Miao, X.; Park, J.A.; Jung, K.-H.; Ahmad, M.Y.; Yue, H.; Ho, S.L.; Liu, S.; Jang, Y.J.; et al. Synthesis, characterization, and X-ray attenuation properties of polyacrylic acid-coated ultrasmall heavy metal oxide (Bi_2O_3, Yb_2O_3, $NaTaO_3$, Dy_2O_3, and Gd_2O_3) nanoparticles as potential CT contrast agents. *Colloids Surf. A* **2019**, *576*, 73–81. [CrossRef]
99. Brown, A.L.; Naha, P.C.; Benavides-Montes, V.; Litt, H.I.; Goforth, A.M.; Cormode, D.P. Synthesis, X-ray Opacity, and Biological Compatibility of Ultra-High Payload Elemental Bismuth Nanoparticle X-ray Contrast Agents. *Chem. Mater.* **2014**, *26*, 2266–2274. [CrossRef]
100. Dadashi, S.; Poursalehi, R.; Delavari, H. Optical and structural properties of oxidation resistant colloidal bismuth/gold nanocomposite: An efficient nanoparticles based contrast agent for X-ray computed tomography. *J. Mol. Liq.* **2018**, *254*, 12–19. [CrossRef]
101. Hu, X.; Sun, J.; Li, F.; Li, R.; Wu, J.; He, J.; Wang, N.; Liu, J.; Wang, S.; Zhou, F.; et al. Renal-Clearable Hollow Bismuth Subcarbonate Nanotubes for Tumor Targeted Computed Tomography Imaging and Chemoradiotherapy. *Nano Lett.* **2018**, *18*, 1196–1204. [CrossRef]
102. Naha, P.C.; Al Zaki, A.; Hecht, E.; Chorny, M.; Chhour, P.; Blankemeyer, E.; Yates, D.M.; Witschey, W.R.T.; Litt, H.I.; Tsourkas, A.; et al. Dextran coated bismuth–iron oxide nanohybrid contrast agents for computed tomography and magnetic resonance imaging. *J. Mater. Chem. B* **2014**, *2*, 8239–8248. [CrossRef]
103. Rivera, E.J.; Tran, L.A.; Hernández-Rivera, M.; Yoon, D.; Mikos, A.G.; Rusakova, I.A.; Cheong, B.Y.; Cabreira-Hansen, M.d.G.; Willerson, J.T.; Perin, E.C.; et al. Bismuth@US-tubes as a potential contrast agent for X-ray imaging applications. *J. Mater. Chem. B* **2013**, *1*, 4792–4800. [CrossRef]
104. Hsu, C.-L.; Lien, C.-W.; Wang, C.-W.; Harroun, S.G.; Huang, C.-C.; Chang, H.-T. Immobilization of aptamer-modified gold nanoparticles on BiOCl nanosheets: Tunable peroxidase-like activity by protein recognition. *Biosens. Bioelectron.* **2016**, *75*, 181–187. [CrossRef]

105. Pene, F.; Courtine, E.; Cariou, A.; Mira, J.-P. Toward theragnostics. *Crit. Care Med.* **2009**, *37*, S50–S58. [CrossRef]
106. Li, Z.; Hu, Y.; Miao, Z.; Xu, H.; Li, C.; Zhao, Y.; Li, Z.; Chang, M.; Ma, Z.; Sun, Y. Dual-stimuli responsive bismuth nanoraspberries for multimodal imaging and combined cancer therapy. *Nano Lett.* **2018**, *18*, 6778–6788. [CrossRef] [PubMed]
107. Xuan, Y.; Song, X.-L.; Yang, X.-Q.; Zhang, R.-Y.; Song, Z.-Y.; Zhao, D.-H.; Hou, X.-L.; An, J.; Zhang, X.-S.; Zhao, Y.-D. Bismuth particles imbedded degradable nanohydrogel prepared by one-step method for tumor dual-mode imaging and chemo-photothermal combined therapy. *Chem. Eng. J.* **2019**, *375*, 122000. [CrossRef]
108. Yang, C.; Guo, C.; Guo, W.; Zhao, X.; Liu, S.; Han, X. Multifunctional Bismuth Nanoparticles as Theranostic Agent for PA/CT Imaging and NIR Laser-Driven Photothermal Therapy. *ACS Appl. Nano Mater.* **2018**, *1*, 820–830. [CrossRef]
109. Yu, N.; Wang, Z.; Zhang, J.; Liu, Z.; Zhu, B.; Yu, J.; Zhu, M.; Peng, C.; Chen, Z. Thiol-capped Bi nanoparticles as stable and all-in-one type theranostic nanoagents for tumor imaging and thermoradiotherapy. *Biomaterials* **2018**, *161*, 279–291. [CrossRef] [PubMed]
110. Detappe, A.; Thomas, E.; Tibbitt, M.W.; Kunjachan, S.; Zavidij, O.; Parnandi, N.; Reznichenko, E.; Lux, F.; Tillement, O.; Berbeco, R. Ultrasmall Silica-Based Bismuth Gadolinium Nanoparticles for Dual Magnetic Resonance–Computed Tomography Image Guided Radiation Therapy. *Nano Lett.* **2017**, *17*, 1733–1740. [CrossRef]
111. Badrigilan, S.; Shaabani, B.; Gharehaghaji, N.; Mesbahi, A. Iron oxide/bismuth oxide nanocomposites coated by graphene quantum dots: "Three-in-one" theranostic agents for simultaneous CT/MR imaging-guided in vitro photothermal therapy. *Photodiagn. Photodyn. Ther.* **2019**, *25*, 504–514. [CrossRef]

materials MDPI

Review

Polymer-Coated Magnetite Nanoparticles for Protein Immobilization

Kinga Mylkie, Pawel Nowak, Patryk Rybczynski and Marta Ziegler-Borowska *

Faculty of Chemistry, Nicolaus Copernicus University in Torun, Gagarina 7, 87-100 Torun, Poland; kinga.mylkie@o2.pl (K.M.); nowak19981411@wp.pl (P.N.); pat_ryb@doktorant.umk.pl (P.R.)
* Correspondence: martaz@umk.pl

Abstract: Since their discovery, magnetic nanoparticles (MNPs) have become materials with great potential, especially considering the applications of biomedical sciences. A series of works on the preparation, characterization, and application of MNPs has shown that the biological activity of such materials depends on their size, shape, core, and shell nature. Some of the most commonly used MNPs are those based on a magnetite core. On the other hand, synthetic biopolymers are used as a protective surface coating for these nanoparticles. This review describes the advances in the field of polymer-coated MNPs for protein immobilization over the past decade. General methods of MNP preparation and protein immobilization are presented. The most extensive section of this article discusses the latest work on the use of polymer-coated MNPs for the physical and chemical immobilization of three types of proteins: enzymes, antibodies, and serum proteins. Where possible, the effectiveness of the immobilization and the activity and use of the immobilized protein are reported. Finally, the information available in the peer-reviewed literature and the application perspectives for the MNP-immobilized protein systems are summarized as well.

Keywords: magnetite nanoparticles; magnetic nanoparticles; immobilization; protein; polymer coating

Citation: Mylkie, K.; Nowak, P.; Rybczynski, P.; Ziegler-Borowska, M. Polymer-Coated Magnetite Nanoparticles for Protein Immobilization. *Materials* **2021**, *14*, 248. https://doi.org/10.3390/ma14020248

Received: 26 November 2020
Accepted: 28 December 2020
Published: 6 January 2021

1. Introduction

Nanotechnology development and the application of nanomaterials has been a very attractive topic for several years. Among synthesized and used nano-sized structures, magnetic nanoparticles (MNPs) are one of the most interesting and widely applied [1,2]. Such nanostructures are usually composed of a magnetic core and the surrounding shell. The core is mostly iron oxides, which can be in the form of magnetite (Fe_3O_4), hematite (α-Fe_2O_3), maghemite (γ-Fe_2O_3), and FeO [3–6]. However, of the mentioned iron oxides, magnetite (Fe_3O_4) is most often used as the core for MNPs preparation [7]. Apart from superparamagnetism, magnetite nanoparticles are characterized by many useful and unique properties such as high surface area, large surface-to-volume ratio, and easy separation under an external magnetic field. Due to this properties, they are widely used in bioseparation, catalysis, industrial processes and, above all, in biomedical sciences [8–11].

The biomedical applications of magnetite nanoparticles are primarily cancer diagnostics and therapies (Magnetic Resonance Imaging, Hyperthermia, Magnetic Field-Assisted Radiotherapy, Photodynamic Therapy), biocatalysis, pharmaceutical analysis, tissue engineering, biosensor, and the immobilization of biomolecules such as proteins [3,12–15]. However, pure, uncoated magnetite nanoparticles have some limitations in use by reason of the ability to spontaneously form aggregates (a result of the system's desire to reduce surface energy, both under the influence of the magnetic field and the biological environment) [16]. Moreover, non-functionalized Fe_3O_4 nanoparticles are characterized by high chemical activity and susceptibility to oxidation, which often leads to a decrease or complete loss of magnetic properties. Covering the magnetite core with a coating, which may be inorganic and organic compounds, surfactants and polymers lead to the stabilization of the nanomagnetic material and, as a result, to the possibility of its wider use, e.g., in industry [17–19].

One of the most universal and widely used applications of magnetite nanoparticles is their treatment as carriers for protein immobilization [20,21]. The main problem in working with free, non-immobilized protein is the difficulty of its separation from the supernatant, which requires such techniques as ultrafiltration [22], ultracentrifugation, and microdialysis [23]. Protein immobilization on a magnetic carrier allows for a simple and quick separation of the nanoparticle–protein system from the supernatant by applying a simple magnet [20]. The basic condition that must be met by such a system is to maintain the activity of the immobilized protein. Moreover, the immobilization of a protein causes its structure to be targeted and, in some cases, this may facilitate e.g., the formation of an active protein–ligand complex. In the case of the immobilization of enzymes on the surface of magnetite nanoparticles, it is possible to reuse such a catalytic system, which is very beneficial [24–27].

As already mentioned, the surface of the magnetite core can be coated with various compounds, of which polymers seem to be the most attractive [12]. Deposition of the polymer on the surface of magnetite nanoparticles provides its chemical and thermal stability [16,28,29]. In addition, in the case of nanoparticles coated with polymers, a smaller tendency to their aggregation is observed, which significantly increases their usability [30,31]. Furthermore, most polymers, and in particular biopolymers, have in their structure reactive amino, carboxyl, and hydroxyl groups, which can be successfully used in the immobilization of proteins [30]. Therefore, in recent years, there has been a growing interest in the synthesis of magnetite nanoparticles coated with macromolecules and their subsequent application for the immobilization of bioligands [32–34].

The following review describes the utility of magnetite nanoparticles coated with polymers in protein immobilization. Both the cases in which the protein was immobilized by physical interactions (adsorption) and by covalent bonding with the polymer surrounding the magnetite core are discussed. Particular attention is paid to nanoparticles coated with natural polymers as those that find wider application in biomedicine as a result of their biocompatibility.

2. Synthesis of Magnetite Nanoparticles—Chemical Methods

Synthetic methods for magnetic nanoparticles preparation can be classified according to the type of process that results in obtaining magnetic nanomaterial. Currently, the following synthesis approaches can be distinguished: chemical, physical, and biological. All of the listed methods have their advantages and disadvantages. Biological methods involving the use of microorganisms for the synthesis of nanoparticles allow for precise control of the obtained material. They are characterized by high process efficiency and very high repeatability. Moreover, they are inexpensive, but the fermentation process is quite time-consuming. Although physical methods for obtaining MNPs are easy to perform, it is difficult to control the size of the resulting nanoparticles in contrast to chemical methods covering about 90% of all techniques (Figure 1) [19].

The first literature reports of MNPs preparation appeared in the late 1960s in the works of Papell, who obtained nanoparticles by the physical method—grinding microcrystalline magnetite powder in the presence of solvents and surfactants in a ball mill. Unfortunately, MNPs obtained by this way were characterized by a high polydispersity index [35]. To prevent particle–particle agglomeration or sedimentation, Papell finally added oleic acid as a dispersing agent [36]. Subsequent works on this subject published by Sugimoto and Matijević concerned the chemical preparation of monodisperse magnetic nanomaterial of narrow size distribution, with mean diameters ranging between 0.03 and 1.1 μm. MNPs were obtained by $FeSO_4$ interaction with potassium hydroxide in the presence of nitrate ion, and the resulting gelatinous suspension was kept at 90 °C for several hours [37]. In 2002, Sun and Zeng developed a synthesis of monodisperse magnetic nanoparticles with a small diameter of 4 nm. They used for this purpose the thermal decomposition of iron (III) acetylacetonate in phenyl ether in the presence of ethanol, oleic acid, and oleylamine [38]. Since then, new methods of preparation of MNPs with the desired sizes and properties are practically constantly being sought.

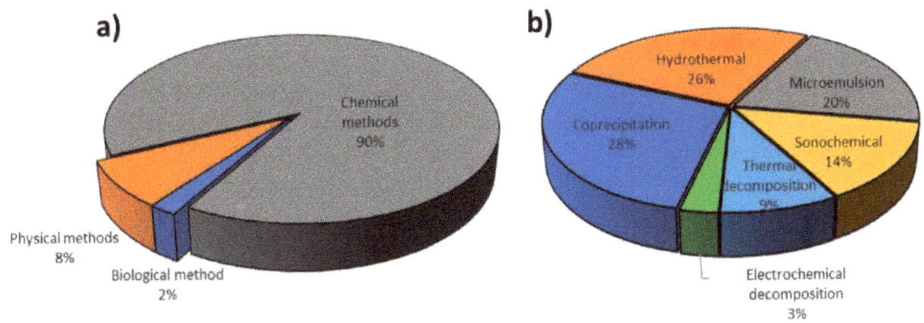

Figure 1. (**a**) A comparison of the synthesis of magnetic nanoparticles (MNPs) by three different routes; (**b**) classification of chemical methods for the synthesis of magnetic nanoparticles.

Due to the fact that the methods of chemical synthesis of magnetic nanoparticles are the most commonly used in their preparation, only these methods are discussed in the next part of this review.

2.1. Co-Precipitation Reaction

The co-precipitation reaction is the most popular method of magnetic nanomaterials synthesis. This route is widely used for the preparation of magnetite nanomaterials intended for biomedical applications due to non-toxic reagents. Co-precipitation is about reaction of magnetite formation in alkaline solution. It is necessary to keep anaerobic reaction conditions to prevent the conversion of magnetite to iron (III) oxide and then iron (III) hydroxide. Such conversion is very unfavorable for biomedical applications due to the ability of iron (III) oxide to form free radicals and the insolubility of iron (III) hydroxide. To avoid the oxidation of magnetite during the synthesis of nanoparticles, various types of stabilizers are introduced to protect the magnetite against oxygen. Moreover, magnetic phase and particle size can be controlled by changing the Fe^{2+}/Fe^{3+} ions ratio, temperature, pH, and the type of used base solution [39,40]. The scheme of the magnetite synthesis from alkaline solutions of iron (II) and (III) salts is presented in Figure 2.

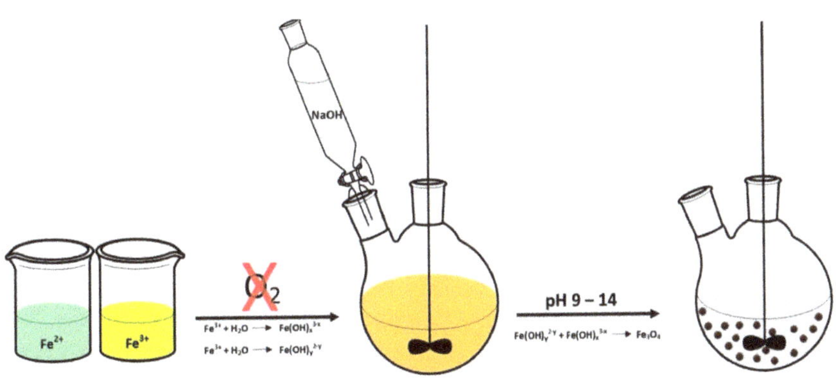

Figure 2. Synthesis of iron oxide nanoparticles by co-precipitation.

2.2. Thermal Decomposition

Thermal decomposition is based on the decomposition of metal precursors (e.g., acetylacetonates, carbonyls, or oleates) at high temperatures (150–300 °C) in the presence of organic solvents characterized with high boiling point (250–300 °C) such as octadecene or benzyl ether [41]. The presence of dispersants and hydrophobic ligands, including oleic acid, lauric acid, oleylamine, and hexadecyl amine, is required to control the size and shape of formed nanoparticles and prevent their aggregation [42,43]. This synthesis procedure is divided into three main steps. Firstly, a mixture of solvent with organometallic precursors, surfactants, and stabilizing agents is heated at a constant rate to reach the nucleation or decomposition temperature of the precursor. Next, the solution is heated to the boiling point of the solvent, which leads to the formation of small nanocrystals, and the final step consists of growth phase in which the solution is refluxed for some time and cooled to room temperature [44]. The schematic representation of this type of synthesis is presented in Figure 3.

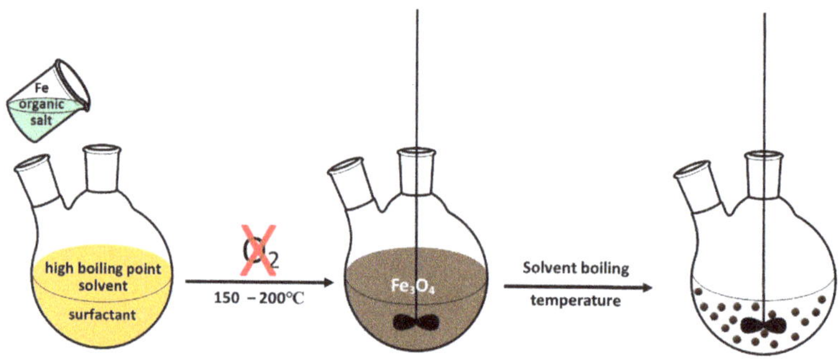

Figure 3. Synthesis of magnetite nanoparticles by thermal decomposition.

However, the synthesis procedure becomes more complex, time-consuming, and more difficult to scale than the co-precipitation reaction; this method allows obtaining magnetite nanoparticles with very narrow size distribution and well-defined magnetic property [16]. Additionally, the appropriate proportions of the reagents and temperature control determine the synthesis of MNPs of defined shapes, sizes, and crystallinity. Unfortunately, there is often a lack of repeatability of the results obtained in this method due to the need to control many parameters during the process. Moreover, the MNPs prepared in this way are dispersible only in non-polar solvents (e.g., hexane) and are not miscible with water (as opposed to those obtained by co-precipitation), which can be a major limitation in some applications, especially biomedical [45].

2.3. Microemulsion Method

Microemulsions are thermodynamically stable colloidal suspensions in which, thanks to surfactants, two initially immiscible liquids coexist in one phase. In this technique, a stable dispersion of two immiscible solvents (water/oil) stabilized with a surfactant (e.g., dodecyl sulfate) is prepared [46]. Substrates for the production of nanoparticles are closed inside the micelle formed from water droplets surrounded by surfactant. The continuous phase in this case is the oil phase. The synthesis of MNPs takes place by introducing a second emulsion or a solution of a base into the system, which causes the rupture and coalescence of the micelles with the substrates, leading to the preparation of magnetite nanoparticles. The use of micelles allows for the stabilization of the system by limiting the nucleation, growth, and agglomeration of nanoparticles [47]. Obtained nanoparticles are separated from the emulsion by extraction with an organic solvent. The microemulsion method leads to nanoparticles with sizes from 1 to 100 nm. Compared to other methods, this procedure has several advantages, including the use of simple apparatus, the ability to synthesize many different materials with a high degree of control of particle size and composition, the preparation of nanoparticles with a crystalline structure and large specific surface area, and the use of simple synthesis conditions closed to ambient temperature and pressure [18]. Properties of magnetic nanomaterial prepared with the microemulsion method depend on the type and structure of the used surfactant [48].

2.4. *Hydrothermal Method*

The hydrothermal method was first described by the German geologist Karl Emil von Schafhäutl in 1845 [45]. The general system is composed of (solid) metal linoleate, a liquid phase of ethanol–linoleic acid, and a hydrothermal (high temperature and high pressure) water–ethanol solution [18]. Typically, the reaction is carried out at about 220 °C, 13.79 MPa pressure for 72 h. [45,49]. As can be deduced from the high-pressure conditions, hydrothermal reactions are carried out using autoclaves or special reactors. For example, Wang, Zhuang, and Peng obtained monodispersed Fe_3O_4 nanoparticles with a size of 9 nm using a vigorously stirred mixture of $FeCl_3$, ethylene glycol (high-boiling reducing agent), sodium acetate (electrostatic stabilizer), and polyethylene glycol as surfactant. The mixture was closed in an autoclave made of Teflon-coated stainless steel at a temperature of 200 °C up to 72 h (Figure 4) [50].

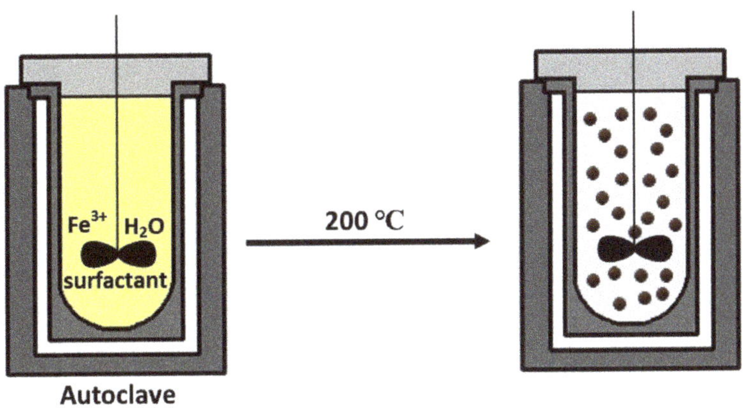

Figure 4. Schematic synthesis of magnetite nanoparticles by the hydrothermal method.

2.5. *Sonochemical Processing*

Sonochemical reactions involve the use of ultrasound to synthesize nanomaterials with a controlled size distribution using co-precipitation or decomposition reactions. [51]. This technique uses during synthesis the phenomenon of acoustic cavitation consisting of the propagation of sound waves in the range from 20 to 10 MHz. Sound waves cause pressure changes in the liquid layer close to the surface, as a result of which gas bubbles form. After placing the solution in an ultrasonic reactor of high temperature (>500 K), high pressure (>20 MPa), and high cooling rate (10^{10} Ks^{-1}), nanoparticles are generated by the collapse of the bubbles [52,53]. These conditions are favorable for the production of highly monodisperse nanomaterials. This method has been used for preparation of several nanocomposites, and its versatility has been successfully confirmed in the synthesis of magnetite nanoparticles [54]. Vijayakumar described the sonochemical synthesis of pure Fe_3O_4 powder, which is a monodisperse material with a particle size of 10 nm, minimizing the nanoparticle agglomeration process [55].

2.6. *Electrochemical Methods*

The literature describes many methods of electrochemical preparation of magnetic nanoparticles [4,56–63]. Moreover, a lot of articles indicate a significant impact of the density (J) and potential (E) of used redox systems on the prepared MNPs size as well as the possibility of its control [57,58,61–63]. In one of them, magnetite nanoparticles, ranging in size from 20 to 30 nm, were produced by electro-oxidation using iron as an electrode material: anode and cathode. The required distance between the two electrodes was about 1 cm. Tetramethylammonium chloride ($(CH_3)_4NCl$) was added to the electrolyte solution as a surfactant, and the reaction was carried out in 30 min at relatively low temperature (6 °C) [57]. This reaction conditions lead to magnetite nanoparticles homogeneous in size and spherical in shape. Subsequently, another electrochemical method was proposed with iron as the anode and water as the electrolyte without any surfactant addition [61]. In this procedure, a change in the distance between the electrodes (from 2 to 6 cm) and the current density from 205 to 415 $\mu A/cm^2$ was used for the size of the resulting nanoparticles regulation. Moreover, it was noticed that the particle size (from 10 to 30 nm) increased simultaneously with the increase of the current density and decrease of the distance between the electrodes [61]. Marques et al. proposed the electro precipitation of magnetic nanoparticles in ethanol solution with iron (III) nitrate nonahydrate ($Fe(NO_3)_3 \times 9H_2O$) as an iron precursor, and the precipitation of $Fe(OH)_3$, which was then reduced to magnetite in the presence of hydroxyl ions formed at the cathode [62]. This route enables the production of superparamagnetic magnetite nanoparticles with controlled size and dispersion [62].

Summarizing, Table 1 shows a comparison of the advantages and disadvantages of the chemical methods of magnetite nanoparticles preparation that were discussed above.

Table 1. Chemical methods for the synthesis of magnetic nanoparticles; their advantages and disadvantages [19,20,64].

Methods	Advantages	Disadvantages
Chemical Co-precipitation	simple and effective	not suitable for the preparation of high-purity accurate stoichiometric phase
Thermal Decomposition	particle size and shapes are controllable	time-consuming synthesis at high temperatures
Microemulsion	uniform properties	surfactants are difficult to remove; small amount can be synthesized
Hydrothermal	particle size and shapes are easily controllable homogeneity	high pressure and high temperature reaction
Sonochemical	size distribution in narrow particle	mechanism not still understood
Electrochemical	easy to control particle size	reproducibility

3. Modification of Bare Magnetite Nanoparticles

Bare magnetite nanoparticles without surface modification with low- or high-molecular compounds have a limited applications as a result of the spontaneous aggregates formation [65]. In addition, uncoated MNPs are characterized by many unfavorable properties, such as chemical instability, poor biodegradability in the physiological environment, and non-specific interaction with blood serum proteins [66]. Moreover, the lack of a stabilizing coating for the magnetic core makes magnetite susceptible to oxidation with oxygen in the air, which as already mentioned is particularly negative for biomedical applications. The cover layer also reduce the risk of adverse effects of material when nanoparticles are used as therapeutic agent [67,68] and allows for the improvement of nanoparticle dispersion in solutions, limiting their toxicity, as well as improving the physicochemical and functional properties. The shell covering the magnetic core can be generated in situ during the preparation of the nanoparticles themselves, or then, the previously obtained pure nanoparticles can be coated. Organic surfactants and inorganic compounds, bioactive compounds, as well as natural and synthetic polymers are used for surface of nanoparticles coating (Figure 5) [69].

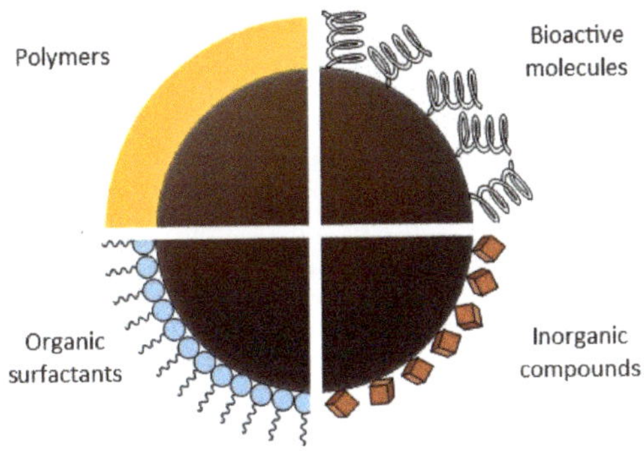

Figure 5. Schematic illustration of the main shells for functionalization of iron oxide magnetic nanoparticles (MNPs).

3.1. *Organic Surfactants*

Surfactants such as oleic acid, lauric acid, alkylsulfonic, and alkylphosphonic acids are used for magnetic nanoparticles stabilization [20]. Sahoo et al. proved based on photos from an electron microscope that carboxylate surfactants provide particles with better isolation and dispersibility than phosphonate surfactants [70]. Unfortunately, long hydrocarbon chains from the surfactant structure cause the nanoparticles to be highly hydrophobic, which makes their use in vivo much more difficult [71]. The research of Luchini and co-authors was aimed at overcoming these limitations. They showed that the use of surfactants with low critical micelle concentration (CMC) values for the functionalization of MNPs resulted in their greater dispersion in solutions, and additionally, the coating process of nanoparticles was more efficient than with surfactants with higher CMC values (Figure 6) [72].

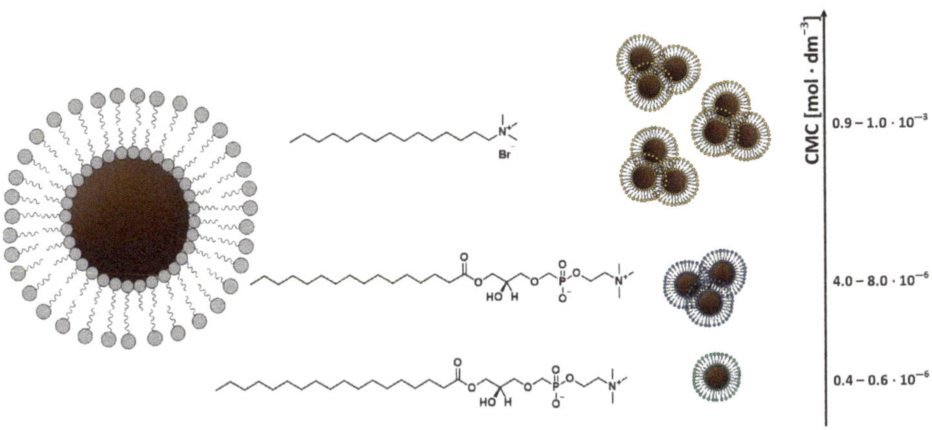

Figure 6. Schematic representation of MNPs and surfactant structures. The clustering tendency is represented as function of values of the critical concentrations.

3.2. *Inorganic Compounds*

Coating the magnetite core with inorganic compounds, apart from ensuring the stability of nanoparticles, quite significantly broadens the area of their application [73]. The most commonly used compounds for this purpose are silicon compounds, metals and their oxides, sulfides, carbon in the form of graphene and its oxide, and nanotubes [74–76].

Silica is a classic coating material for nanoparticles, as it can enhance their dispersion in aqueous solutions, endows them greater durability, and protects them in acidic environments [69,77,78]. The coating of nanoparticles with silica (SiO_2) was carried out based on a modified Stöber process [79]. This method uses tetraethoxysilane (TEOS) as a silicon shell precursor in the hydrolysis and condensation reaction (Figure 7). The synthesis is carried out in an aqueous ethanol solution in the presence of ammonia as a homogeneous catalyst [79]. The presence of silanol groups on the surface of magnetic nanoparticles reduces their potential toxicity and also causes the colloidal stability of nanoparticles in the physiological pH range. Thanks to that, MNPs can be successfully used in molecular biology and medicine [80–82].

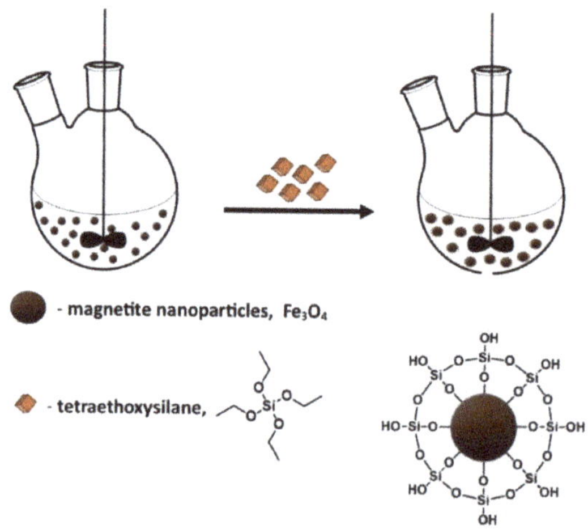

Figure 7. Functionalization of the surface of MNPs using tetraethoxysilane (TEOS).

Another example of a silicon compound used for magnetite core stabilization is aminosilane (SiO_2-NH_2). This compound due to the presence of the basic amino group is positively charged at a pH value of about 7.4. Since plasma membranes have huge negatively charged domains, it has been proved that cationic aminosilan easily penetrates inside the cells [83,84].

Carbon-based materials as an inorganic compound are also applied to MNPs surface coatings to enhance their stability, biocompatibility, and dispersity. Carbon-coated magnetite nanocomposites have found wide application as catalysts, electrode supercapacitors, microwave absorbers, and anode materials for lithium-ion batteries [74–76].

Noble metals such as gold and silver, which are characterized with biocompatibility and resistant to chemical reactions such as oxidation and corrosion, were usually used for MNPs stabilization [85]. The surface of the nanoparticles covered with these metals characterizes not only the stability under physiological conditions, but also, its ligand binding ability would be enhanced and also the formation of harmful free radicals could be prevented [86].

Increasingly, metal oxides such as TiO_2 [87–89], SnO_2 [73,90], Cu_2O [91], ZnO [92], CdS [93], ZnS [94], PbS [95], Bi_2S_3 [96], or sulfides have been also used for the protection or functionalization of MNPs. For example, Saffari et al. prepared superparamagnetic Fe_3O_4-ZnO nanocomposites with 10% ZnO content by adopting the sonochemical method. It was reported that the Fe_3O_4/ZnO nanocomposite has excellent photocatalytic properties. [92].

3.3. Polymers and Bioactive Molecules

Biological active compounds such as lipids, peptides, and proteins are used for magnetic nanoparticles coating without loss of material magnetization [73,80,97,98]. Jahanban-

Esfahlan and co-authors deposited human (HSA) and bovine (BSA) serum albumin on magnetic nanoparticles surface via desolvation [99]. Such nanoparticles have a negatively charged surface that prevents electrostatic interactions with negative biological elements such as plasma proteins and blood cells, thus preserving the stability of nanoparticles [100]. Nosrati used a magnetic nanoparticles coated with BSA prepared by chemical desolvation and co-precipitation as curcumin carriers [100].

Polymers are widely used as a shell covered the surfaces of nanomaterials. They can prevent MNPs oxidation and give nanoparticles collateral stability [101]. Several approaches have been developed for MNPs functionalization with polymers, where the common methods include in situ and post-synthesis coating [73]. In the in situ approach, the conventional routes are mini/micro-emulsion polymerization and the sol–gel process, while the post-synthesis coating method is carried out as a result of chemical reactions or by non-covalent interactions of the polymer with the magnetite core. [73]. The nature of the polymer may be synthetic: polyethylene glycol, polyacrylic acid, poly (vinylpyrrolidone), poly (vinyl alcohol), poly (methacrylic acid) or natural, as in the case of chitosan, starch, cellulose, agarose, and dextran.

4. Protein Immobilization Methods

Protein immobilization is a biotechnological technique where protein is fixed in a suitable matrix that restricts its movement to increase stability and, in the case of an enzyme protein, to allow its reuse with maintaining immobilized protein activity [102]. The first literature reports on the immobilization of catalytic proteins come from 1916 when Griffin and co-authors adsorbed the enzyme invertase on a solid matrix of charcoal and aluminum hydroxide [103].

Choosing the right immobilization technique for a given protein plays a very important role. The cost of immobilization, the possible inactivation of the protein, the toxicity of the reagents, and the properties of the system obtained in this way should be taken into account [104]. Protein immobilization methods can be divided into two categories: chemical and physical. The chemical method uses the formation of new covalent or ionic bonds, and the physical method mainly involves hydrophobic or van der Waals interactions between the protein and the support. While chemical immobilization creates strong bonds between the molecule and the carrier, this results in greater durability of the immobilization compared to physical immobilization but may cause changes in the structure of the immobilized molecule. The most commonly used techniques for the immobilization of proteins, including enzymatic proteins, are adsorption, trapping, covalent bonding, and cross-linking. As a rule, immobilized proteins are less sensitive to changes in pH, temperature, and the action of toxins. Due to the possibility of changing the protein structure, immobilization is not a flawless method. For example, carrying out the immobilization process may reduce the catalytic properties of the enzyme. Factors that may affect the activity of the immobilized protein are the stiffening of the protein structure and diffusion resistance in the free transport of ligands to and from the active center of the protein [105].

There are some parameters describing the performed immobilization such as the yield of immobilization (protein loading), and in the case of enzymatic protein, efficiency and activity recovery [106].

In processes using proteins such as analytics, bioseparation, and catalysis, separation of the used protein from the supernatant is a key parameter prompting the use of immobilization. As already mentioned, the immobilization of the protein on the surface of magnetic nanoparticles (MNPs) primarily ensures easy separation of the carrier–protein system using an external magnetic field. In recent years, MNPs are one of the most frequently used supports for this purpose because of their large surface area and easy functionalization. The main advantage of using MNPs as carriers for the immobilization of e.g., enzymatic proteins is the possibility of reusing the biocatalyst after separation from the reaction medium. Immobilization on a magnetic carrier is mainly physical immobilization and covalent bonding on the surface of the support. Due to the fact that the immobilization takes place on the surface of a permanent support, physical adsorption and covalent binding are usually applied [104,107–109].

4.1. Adsorption of Protein on MNPs Surface

As it was discussed, the adsorption of proteins on solid carriers is one of the oldest and simplest physical immobilization techniques. The protein can be immobilized by mixing with an appropriate adsorbent. It is important to ensure the right immobilization conditions: pH and ionic strength of the solution. Only weak interactions, such as hydrophobic and van der Waals interactions and hydrogen bonds, keep the protein molecule on the surface of the carrier. In general, enzyme immobilization through the technique of physical adsorption is quite simple and may have a high commercial potential due to its simplicity, low cost, and retaining high enzyme activity as well as a relatively chemical-free biomolecules binding. Naturally, this method suffers from several disadvantages such as low resistance to changes in pH, temperature, and the ionic strength of the buffer. Furthermore, the physical interaction is generally too weak to hold the protein bound to the carrier, which may cause the protein desorption from the carrier surface in solution [109–111]. Desorption usually leads to a loss of protein activity and contamination of the supernatant with protein, which may prevent the reuse of immobilized proteins, especially for analytical applications. Additionally, protein adsorption on the carrier surface often leads to conformational changes in its structure and loss of protein activity. This immobilization technique allows the use of many carriers, because the most important part of this method is the appropriately high affinity of the carrier to the immobilized protein. However, depending on the type of used carrier, the amount of deposited protein may be different, and the effectiveness of immobilization depends not only on the type of support but also on the used enzyme and the immobilization conditions.

4.2. Covalent Binding of Protein on MNPs Surface

Covalent immobilization is a chemical method of protein binding on the carrier surface. It involves the formation of a covalent bond between the functional groups of the protein and support. It is advisable to use such protein functional groups in the immobilization process, which are susceptible to chemical modification and do not participate in the stabilization of the third and fourth-order structure of the protein [104]. The cysteine thiol group, phenyl ring of tyrosine, imidazole group of histidine, and amino group of lysine are most often used for this purpose. In order to easily react the above-mentioned functional groups of the protein, the surface of the support should contain reactive amino, hydroxyl, carboxyl, vinyl sulfone, vinyl ketone, oxirane, aldehyde, halide, and thiol groups. Depending on the type and presence of functional groups on the carrier surface, the formation of the carrier–protein covalent bond consists in the following reactions: arylation, amidation, diazotization, alkylation, and the formation of Schiff bases or amide bonds. Coupling agents such as glutaraldehyde (GA) [112], glyoxal [4], epichlorohydrin [104], 1-ethyl-3-(3-dimethyl-aminopropyl) carbodiimide (EDC), and *N*-hydroxysuccinimide (NHS) [113] are often used in protein covalent immobilization [114]. For example, EDC activates carboxyl groups of protein and forms an amine reactive O-acylisourea intermediate that spontaneously reacts with primary amines to form an amide bond and an isourea by-product. The O-acylisourea intermediate is unstable in aqueous solutions, and failure to react with an amine will cause hydrolysis of the intermediate, regeneration of the carboxyls, and the release of an *N*-substituted urea [115,116]. Therefore, it is necessary to quench the EDC activation reaction with a thiol-containing compound such as 2-mercaptoethanol. EDC couples NHS to carboxyls, which forms an NHS ester that is considerably more stable than the O-acylisourea intermediate and allows for efficient conjugation to primary amines at physiological pH. The advantageous quality of EDC is that it is water soluble and dissolves in aqueous buffer solutions, similar to most biological macromolecules [117]

The efficiency and effectiveness of immobilization depends, inter alia, on the number of these groups present on the surface of the support. Naturally, the immobilization efficiency increases with the increase in the number of functional groups on the carrier surface that are available for protein binding. However, in the case of the enzyme protein, too much "packing" of the enzyme on the support surface may lead to decrease in enzymatic activity or inactivation [106].

The number of carriers that can be used in this immobilization technique is less than in physical immobilization. The support surface can be covered with inorganic and organic compounds as well as composites and polymers, and the most important criterion when selecting a carrier is the presence of reactive functional groups on its surface that are able to form covalent bonds.

As it can be seen, each of the methods has its advantages and disadvantages, which are briefly summarized in Table 2.

Table 2. Advantages and disadvantages of most commonly known immobilization methods [114,118,119].

Method of Immobilization	Binding Nature	Advantages	Disadvantages
Adsorption	weak interactions such as hydrogen bond, hydrophobic, and van der Waals interactions	- does not or little affects the enzyme structure - simple, cheap, and easy - no conformational change of the protein - no need to use reagents	- low stability - non-specific adsorption

5. Immobilization of Proteins on Polymer-Coated Nanoparticles

As a result of susceptibility to modification, thermal stability, and resistance to pH changes, as well as mechanical properties, polymers are very often used for a magnetite nanoparticles core coating. As it is known in view of the structure and source of obtaining the polymer, we can divide them into synthetic and natural polymers (biopolymers). Consequently, in this article, polymer-coated nanoparticles used for protein immobilization were divided based on this division.

5.1. *Immobilization of Proteins on Nanoparticles Coated with Synthetic Polymers*

Synthetic polymers, in contrast to natural macromolecules, are prepared from the corresponding monomers by chemical synthesis. These polymers are characterized by high purity and their weight and composition are controlled in the synthesis process. The purity of the material is of particular importance for their use in biomedicine and e.g., pharmaceutical formulation [120]. The basic methods of obtaining synthetic polymers are generally bulk, solution, suspension, or emulsion technology using homogeneous or heterogeneous catalysts of acid, alkali, or radical species or transition and rare metal catalysts. Moreover, synthetic polymers are versatile materials that can be processed into biomedical foams with a wide range of mechanical, thermal, and degradation properties. The tailoring of these properties can be achieved by using different polymeric families such as polyesters, polyurethanes, and tyrosine-derived polymers. One of the parameters determining the final properties of a polymer is also its porosity [121]. The applications of magnetite nanoparticles coated with synthetic polymers in protein immobilization are presented below.

5.1.1. Immobilization of Proteins on Nanoparticles Coated with Polyethylene Glycol (PEG)

Polyethylene glycol (PEG) is hydrophilic, uncharged, and non-immunogenic linear polyether macromolecule. As a result of its low toxicity and ease of excretion from the body through the kidneys (for PEGs less than 30 kDa) and in feces (for PEGs > 20 kDa), PEG is also very often used for biomedical and therapeutic applications [122]. In addition, of all the synthetic polymers, this macromolecule is the most widely used for magnetic core coating, especially to ensure the high colloidal stability of the nanomaterial [123].

However, in the literature, there are only a few reports on the use of PEG-coated magnetite nanoparticles as a carrier for proteins.

On the basis of Mukhopadhyay's [124] research, it can be concluded that PEG coatings are effective means of protecting biomolecules against the toxicity generated by magnetite nanoparticles. Mukhopadhyay [124] et al. described the interaction of magnetic nanoparticles coated with ethylene glycol and Cytochrome C, which acts as an electron transporter in the respiratory chain. The interaction of naked uncoated nanoparticles with Cytochrome C led to the reduction of this protein, while the magnetite nanoparticles coated with PEG showed no affinity for Cytochrome C and finally did not reduce it (Figure 8).

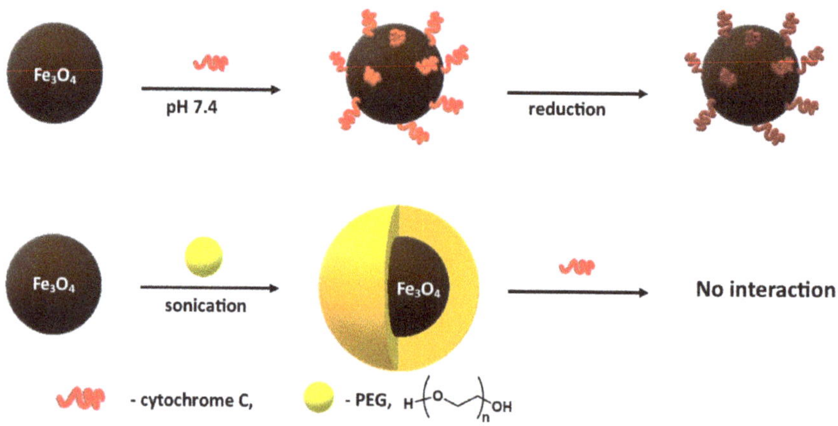

Figure 8. Schematic representation of the interaction of Cytochrome C with the bare and polyethylene glycol (PEG)-coated Fe_3O_4 MNPs.

Chang and co-authors successfully applied magnetite nanoparticles coated with polyethylene glycol (PEG) to the novel method of the exosome purification. For this purpose, they used nanoparticles obtained by chemical co-precipitation reaction. Next, this material was used to remove fetal bovine serum (FBS) from a biological fluid by means of physical immobilization [125]. Exosomes are secreted e.g., by cancer cells that are responsible for metastasis and subsequent cancer growth. The detection of exosomes is a key step in the early cancer diagnosis [126,127]. Exosomes exist in a biological fluid such as blood that also contains proteins, so it is important to remove proteins from the biological fluid before exome detection to avoid test interference. The research results indicate that the proposed method of using PEG-coated magnetite nanoparticles for protein removal and exosome purification is quick and simple.

5.1.2. Immobilization of Proteins on Nanoparticles Coated with Polyvinyl Alcohol (PVA)

Polyvinyl alcohol (PVA) is hygroscopic, colorless, and odourless biocompatible and biodegradable synthetic macromolecule [128]. One of the most important properties of PVA is the ability to form multiple hydrogen bonds between the polymer chains, which prevents aggregation and agglomeration of magnetic nanoparticles coated with this polymer [80,129,130].

Mahmoudi [131] et al. published the physical immobilization of human protein—transferrin on magnetite nanoparticles, both naked and coated with PVA. In addition, it was also the first article to describe the irreversible conformational changes of a specific protein as a result of interaction with MNPs. After the adsorption of transferrin on the nanoparticles surface, the main function of the protein, which is the transport of iron between cells, was changed. The changes that took place in the structure and activity of transferrin were irreversible. The activity of the desorbed protein was tested, and transferrin was found to be inactive.

Laochai [132] and co-authors published the synthesis of magnetite nanoparticles coated with a mixture of polyvinyl alcohol and chitosan (CS). Bare magnetite nanoparticles were synthesized by a simple method of co-precipitation and in the next stage coated with PVA and s chitosan (CS) layer. Then, the horseradish peroxidase was immobilized on the surface of prepared nanoparticles. This enzyme is usually used for hydrogen peroxide detection via catalytical oxidation of the substrate of hydrogen peroxide and o-dianazidine. A colored product of this reaction is formed whose concentration is proportional to the concentration of enzyme and measured [133]. It was also proved that peroxidase immobilized on PVA/CS-coated magnetic nanoparticles retained its activity in ten catalytic cycles.

Sahin and co-authors published the synthesis of magnetic nanoparticles coated with polyvinyl alcohol and trypsin covalent immobilization on their surface with glutaraldehyde as a linker [134]. Trypsin is hydrolase that selectively catalyzes the hydrolysis of peptide bonds [135]. According to the published results, immobilized trypsin showed at 40 °C greater stability than the free enzyme. Additionally, after 12 days of storage at 4 °C, immobilized enzyme retained about 50% of its initial activity, while the activity of free trypsin stored under the same conditions was only 19%. In addition, the reuse of immobilized enzyme was also tested. After eight catalytic cycles, the trypsin deposited on the MNPs retained 56% of the original system activity. The efficacy of the immobilized trypsin was assessed in a study based on the digestibility of Cytochrome C. Immobilized trypsin showed effective proteolytic activity in a shorter time (15 min) than free trypsin (24 h).

5.1.3. Immobilization of Proteins on Nanoparticles Coated with poly(D,L-lactide-co-glycolide) (PLGA)

Poly(D,L-lactide-*co*-glycolide) is a copolymer synthesized by a random ring-opening copolymerization of two different monomers, cyclic dimers of glycolic acid (1,4-dioxane-2,5-diones) and lactic acid [136]. It shows great potential in drug transport and tissue engineering due to its biocompatibility and biodegradability [137,138], and it is soluble in most commonly used solvents. Depending on the proportion of each monomer in the copolymer structure and polymer molecular weight, PLGA shows different properties. Usually, higher amounts of lactide in PLGA lead to less hydrophilic material with slow degradation as a result of lower water absorption [69].

PLGA-coated magnetite nanoparticles have many advantages in drug delivery, especially in drugs protection against degradation. Moreover, they can also improve the pharmacokinetic and pharmacodynamic profiles of transported drugs. Another important advantage of PLGA over other polymers is that this macromolecule is approved by the FDA (U.S. Food and Drug Administration) and EMA (European Medicines Agency) in various drug delivery systems, which means that PLGA-coated nanoparticles can be used in clinical trials in a shorter period of time [139,140]. One of the most interesting applications of MNPs in biomedicine is their design to have additional fluorescent or luminescent properties. Such structures can be obtained by chemical or physical modification of the polymer shell surrounding the magnetite core [141,142]. This multimodal approach ensures the specific recognition and attachment of MNPs to the target cell, drug delivery, and the possibility of in vitro and in vivo bioimaging using optical methods, tomography, or magnetic resonance.

Jaemoon Yang and co-authors synthesized a multimodal nanocomposite using inorganic and organic materials for cancer detection and treatment [143]. The nanoemulsion method was used to incorporate doxorubicin, an anti-cancer drug (DOX), into magnetic nanoparticles coated with a mixture of poly (D,L-lactide-co-glycolide) and PVA (Figure 9) [144]. The Herceptin (HER) antibody used in the treatment of breast cancer was also immobilized on the surface of these nanoparticles (with EDC/NHS as a linker) without loss of its affinity for cancer cells. Additionally, the drug enclosed in the polymer shell was released in a balanced manner without any inhibition. These multi-functional composites may have applications in targeted drug delivery, MRI probes, and also cell separation.

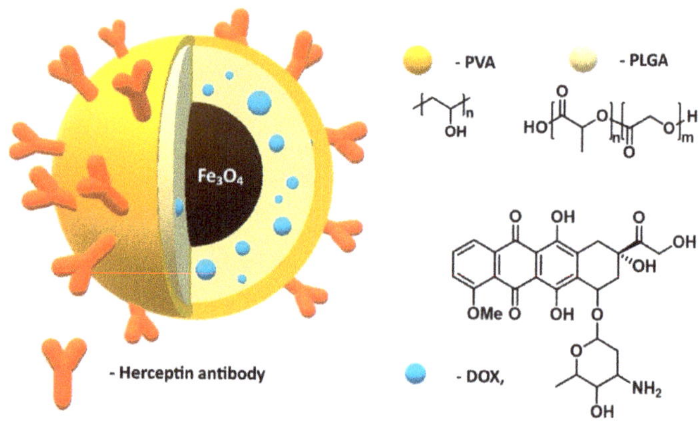

Figure 9. Schematic illustration of magnetic polyoly(D,L-lactide-co-glycolide) (PLGA)/polyvinyl alcohol (PVA)/doxorubicin (DOX) nanoparticles for diagnosis and treatment of cancer.

Shubhra and co-authors published a modification of the surface of magnetite nanoparticles coated with poly (D,L-lactide-co-glycolide) with a non-ionic copolymer—poloxamer (Pluronic F68, PF68) for the immobilization of bovine serum albumin (BSA) [145]. The protein adsorption capacity of this material was compared with poly(D,L-lactide-*co*-glycolide)-coated nanoparticles (Figure 10). The UV-Vis spectrophotometric analysis showed that the BSA adsorption on PF68-modified nanoparticles was reduced by about 50% in relation to poly (D,L-lactide-co-glycolide) modified nanoparticles.

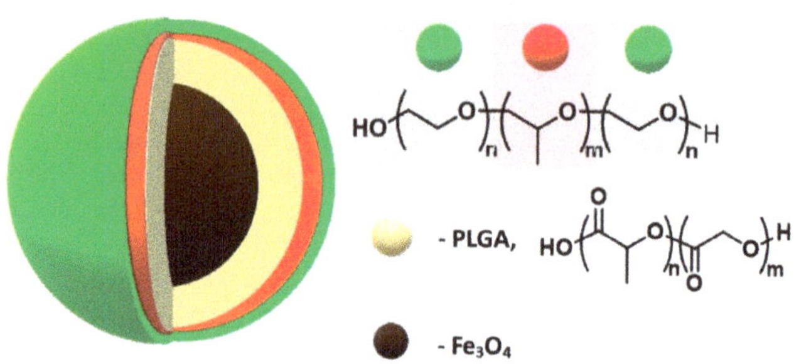

Figure 10. Scheme structure of poloxamer and adsorption of poloxamer onto MNPs-coated PLGA.

5.1.4. Immobilization of Proteins on Nanoparticles Coated with Polyethyleneimine (PEI)

Linear polyethyleneimines contain only secondary amines as opposed to branched PEI, which contains primary, secondary, and tertiary amine groups. There are two carbon atoms in each PEI molecule with one protonated nitrogen atom. Due to the different pK_a values of the primary, secondary, and tertiary amine groups, PEI can scavenge protons under different pH conditions, which is known as the "proton sponge" mechanism. PEI was developed to condense DNA through the electrostatic interaction between the positive and negative charges of the DNA phosphate group [146,147]. A quaternary ammonium derivative of branched polyethyleneimine (bPEI-met) has also been synthesized, and it was found that it exhibits antibacterial activity by disrupting bacterial cell membranes [148]. Due to its unique properties, PEI appears to be one of the most suitable molecules for the surface modification of MNPs for biomedical applications.

In 2016, Xia and co-authors described the synthesis of polyethyleneimine-coated nanoparticles (Fe_3O_4–NH_2–PEI MNPs) [149]. Afterwards, this nanomaterial was chelated with copper ions in order to immobilize the *Trametes versicolor laccase* by physical immobilization—adsorption (Figure 11). Laccase as a multi-copper oxidase (belongs to the group of polyphenol oxidases) can be produced by numerous plants, funguses, and bacteria [150,151]. As a result of its relatively low substrate specificity and high catalytic activity, it has gained extensive attention in various fields such as environmental remediation, the pulp and paper industry, and biosensing [152]. However, the industrial applications of this enzyme are limited due to the low stability and poor reusability of free laccase [150,153]. The results obtained for the enzyme immobilized on (Fe_3O_4–NH_2–PEI MNPs) nanoparticles were compared with those obtained for the enzyme immobilized on the particles without a polymer coating (Fe_3O_4–NH_2 MNPs) containing only amino groups. It was noticed that polymer-modified nanoparticles show a higher adsorption capacity compared to nanoparticles without the polymer coating. Additionally, the recovery of laccase activity for nanoparticles with polyethyleneimine was two times higher than for magnetite nanoparticles without polymer. In addition, the activity of the immobilized enzyme improved significantly; the specific laccase activity was 101.33 times higher than for free enzyme. Moreover, immobilization allowed the enzyme to be reused. Immobilized enzymes on PEI-coated nanoparticles preserved 44.89% of their original activity after the 5th reuse. The activity loss in these steps may be related to particle agglomeration and the inactivation of laccase upon use.

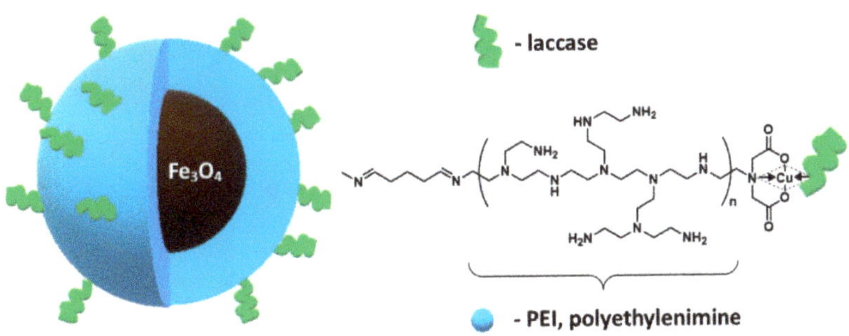

Figure 11. Scheme polyethylenimine-modified Fe_3O_4 nanoparticles (Fe_3O_4–NH_2–PEI NPs) structure with chelated Cu^{2+} and immobilize laccase.

Interesting research with polyethyleneimine-coated magnetic nanoparticles in gene therapy—magnetofection was published by Zuvin et al. in 2019 [154]. In this study, a new magnetic trigger system consisting of four rare earth magnets on a rotating table for a better magnetofection effect was designed and manufactured. Magnetic nanoparticles coated with polyethyleneimine with a green fluorescent protein (GFP) carrying DNA were used as model material. Magnetofection has been tested on the breast cancer cell line (MCF7). The results showed that the magnetic field exposure increased the transfection efficiency.

It can be seen that polymers have different protein adsorption capacity resulting from the presence of different functional groups, different arrangements, and different molecular weight. Wiogo et al. performed the co-precipitation synthesis of bare magnetic nanoparticles that were in the next step modified by sonication with linear polymethacrylic acid (20 kDa), linear and branched polyethyleneamine (25 kDa), and branched oligoethyleneimine (800 Da) [155]. Next, the adsorption capacity of each material was tested using the biological serum proteins (fatal bovine serum). Based on the obtained results, it can be concluded that nanoparticles coated with branched polyethyleneimine adsorb the largest amount of serum protein, while nanoparticles coated with linear polymethacrylic acid showed the lowest interaction with proteins. The differences in the amount of immobilized protein resulted from the conformation of the polymer on the surface of the MNPs. The results of the interaction of nanoparticles obtained by Wiogo et al. with serum proteins are presented in the Table 3.

Table 3. Results of LC-MS/MS analysis checking which of the FBS proteins remain on the surface of bare MNPs and MNPs functionalized with polymethacrylic acid, linear polyethylenimine, and branched polyethylenimine after MNPs have been contacted with biological solution containing 10% FBS [155].

Protein	Bare MNPs	MNPs Coated with Polymethacrylic Acid	MNPs Coated with Linear Polyethylenimine	MNPs Coated with Branched Polyethylenimine
Albumin	+	+	+	+
Antithrombin	-	-	-	+
α-2-HS-glycoprotein	+	-	-	-
Inter- α-inhibitor	-	-	+	+
Apolipoprotein A-1	-	-	+	+
Apolipoprotein E	-	-	+	+
Complement Component 4A	-	-	+	+
Tetranectin	+	+	-	-
α-fetoprotein	-	-	-	+
α-1-antiproteinase	-	-	-	+
Kininogen	-	+	-	-
Complement factor H	+	-	-	-
Hemoglobin	+	+	+	+
Immunoglobulin	-	+	-	-
Complement factor I	+	-	-	-
Complement factor B	+	+	-	-
Apolipoprotein B	+	-	-	-
Lactoferrin	-	+	-	-

It was shown that the design of nanoparticles for biomedical applications can be improved through the appropriate selection of functionalization polymers and understanding the factors governing the stabilization mechanism.

Kannan et al. reported the immobilization of two lipases: *Candida rugosa* and *Mucor miehei* on polyethyleneimine (PEI)-coated MNPs [156]. These nanoparticles were used in chromatography as an anion exchanger for lipase separation, which resulted in retaining a significant part of the enzyme on the support. Moreover, PVP (polyvinylpyrrolidone) has been used in this research to reduce the amount of protein bound to the filling. The PEI-coated Fe_3O_4 nanoparticles were further coated with (1.0–2.5%) PVP solution at about 4 °C with 200 rpm for overnight to (1.0–2.5%) PVP-PEI-Fe_3O_4. Shielding with 2% PVP improved the elution of lipases with 1 M NaCl as eluent. The elution of *Mucor miehei* lipase increased from 56.8% to 68.3%. Similar results were observed for lipase from *Candida rugosa*.

Gräfe et al. [157] and Calatayud et al. [158] published the formation of the protein corona on PEI-coated magnetic nanoparticles. Materials designed for biomedical and therapeutic applications generally come into contact with protein reach body fluids. Proteins are usually adsorbed on the surface of material to form an enveloping layer known as the "protein corona". This process can be thought of as physical immobilization. The protein corona, which is formed as soon as nanoparticles come into contact with biological systems, plays a key role in the biological role of nanoparticles. Gräfe et al. describe a strategy to control the amount of adsorbed proteins on the surface of nanoparticles and the effect of such corona proteins on particle–cell interactions. Polyethyleneimine (PEI)-coated magnetic nanoparticles (MNPs) were incubated in a medium consisting of fetal calf serum (FCS) and the nutrient broth used for cell culture. As it can be expected, during the incubation process, the surface of PEI-modified magnetic nanoparticles was covered with serum proteins by physical immobilization. Next, the human HBMEC (Human Brain Microvascular Endothelial Cells) line was used to study the interaction with nanoparticles.

The results show that the presence of the corona reduces the interaction of the nanoparticles with HBMEC during short-term incubation depending on FCS concentration [157].

On the other hand, Calatayud et al. [158] focused on the process of protein adsorption on magnetic nanoparticles functionalized with polyethyleneimine and poly (acrylic acid) after immersing them in a cell culture medium (Figure 12). It was noticed that after 24-h incubation, large aggregates of proteins are formed on MNPs with a hydrodynamic size of 1500 nm (for nanoparticles coated with poly aryl acid) and 3000 nm (for nanoparticles coated with polyethyleneimine) [158].

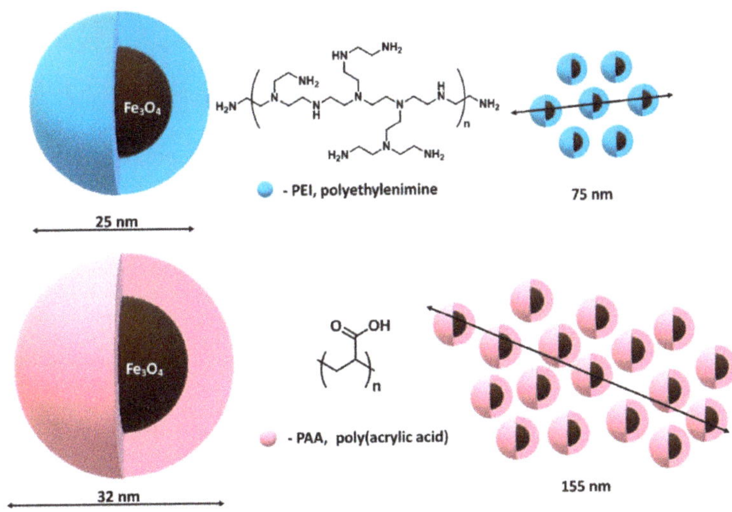

Figure 12. Sketched evolution of the particle agglomeration process for the MNPs when in their as-prepared suspension in water.

The study presents the effect of the produced protein clusters properties on the absorption of SH-SY5Y cells. Despite the negative z-potential with similar values for both MNPs in cell culture, it was observed that PEI-MNPs are incorporated in in much greater amounts than polyacrylic acid (PAA)-MNP units. Quantitative analysis showed that SH-SY5Y cells can incorporate 100% of the added PEI-MNPs up to 100 pg/cell, whereas for PAA-MNPs, the uptake was less than 50%. This result suggests the possibility of controlling non-specific protein adsorption onto MNPs by proper functionalization of their surface. The impact of the final properties of these clusters on the cell uptake, which is typified by the much larger mass of attached PEI-MNPs compared to the PAA-MNPs, was demonstrated.

5.1.5. Immobilization of Proteins on Nanoparticles Coated with Polyacrylic Acid (PAA)

Polyacrylic acid (PAA) is a non-toxic polyanion (in which each unit has a carboxyl group), synthetic, high molecular weight polymer [159]. PAA is a weak polyelectrolyte with a degree of dissociation depending on the pH of the solution and its ionic strength. Moreover, it is a water-soluble macromolecule with a high density of reactive functional groups. These properties provide strong connections between iron oxide and biomolecules, which makes PAA-coated magnetite nanoparticles very attractive for biomedical applications [160]. Based on Sanchez's observations, it is known that the amount of PAA covering the magnetic core affects the size and polydispersity of nanoparticles: the greater the proportion of PAA, the smaller and more monodisperse the nanoparticles that were obtained [161].

Hamidrez and co-authors modified magnetic nanoparticles coated with a mixture of PAA and chitosan with two proteins: BSA (Bovine Serum Albumin) and IgG (Immunoglobulin G). The surface of the nanoparticles was first coated with chitosan in order to stabilize the structure; then, polyacrylic acid (PAA) was used as the outer layer [162]. After incubation with BSA and IgG protein, it was noted that the adsorption of BSA was very low in contrast to IgG protein binding. These results indicate that magnetic nanoparticles coated with polyacrylic acid (PAA) and chitosan mixture can be a good carrier for the transport of drugs [163].

An interesting application of magnetic nanoparticles coated with polyacrylic acid (PAA) was described by Huang et al. [164]. PAA-coated magnetite nanoparticles were obtained by co-precipitation and then used for the immobilization of the lipase from *Candida rugosa* from aqueous solutions [165]. It was shown that the maximum lipase adsorption was found to be 0.605 mg of enzyme per 1 mg of nanoparticles. Additionally, the desorption process and the enzyme activity after this process were investigated. The percentage of enzyme desorption was about 80%, while the enzyme activity recovery was 95.5% of the initial lipase activity value before the immobilization process [164].

Ma et al. used polyacrylic acid (PAA)-coated magnetic nanoparticles in the treatment of arterial embolism [166]. Tissue plasminogen activator (rtPA), an enzyme responsible for the fibrinolysis process, was immobilized on such surface-modified nanoparticles, using the amide bond formed with EDC/NHS. The system prepared by this way was used in an animal model to test its effectiveness in the physiological process of breaking down an arterial thrombus. The rtPA activity tested after immobilization was about 87% of the initial enzyme activity. Moreover, the results described in the conclusions indicate that after 75 min, it was possible to improve the patency of the artery to 82%, and the used material did not adversely affect the number of blood cells and hemoglobin [166].

In the case of the design and synthesis of materials as drug carriers, it is important to study their interaction with serum proteins. Zhao and co-authors [167] studied the interactions of lysozyme (LYZ) and bovine serum albumin (BSA) with two types of magnetic nanoparticles: coated with polyacrylic acid (PAA) and the other type resulting from the modification of PAA-coated MNPs with 3-(diethylamino) propylamine (DEAPA) (Figure 13) [168]. Poly(acrylic acid) modified with (PAA)-co-3-(diethylamino)-propylamine (DEAPA)) is an important zwitterionic polymer with positively and negatively charged moieties on different monomer units.

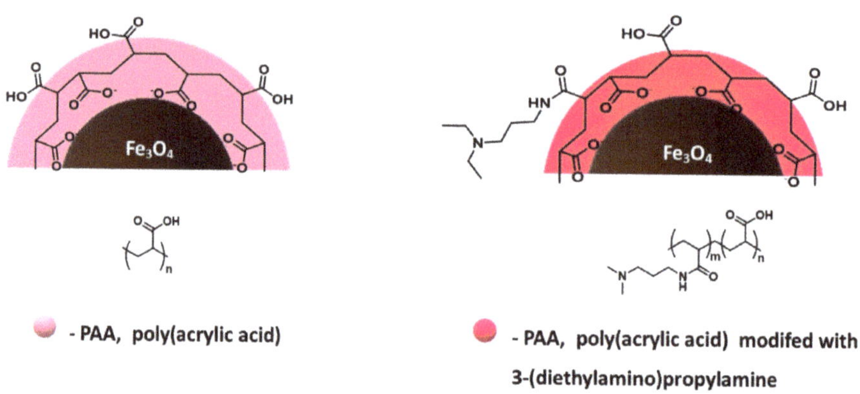

Figure 13. Magnetite nanoparticles coated with polyacrylic acid (PAA) and 3-(diethylamino)propylamine-modified PAA.

The results of the study of the interaction of these nanoparticles with LYZ and BSA showed that the lysozyme has a high affinity for nanoparticles coated with pure, unmodified PAA, in contrast to BSA, which deposited small amounts on this material. For nanoparticles coated with PAA modified with 3-(diethylamino) propylamine, it was observed that both proteins LYZ and BSA were adsorbed in significant amounts, which was higher than for nanoparticles coated with PAA alone [167].

5.1.6. Immobilization of Proteins on Nanoparticles Coated with Poly(methacrylic acid) (PMAA)

Poly(methacrylic acid) (PMAA) is a polymer synthesized from methacrylic acid, which is a viscous liquid with a specific smell. PMAA has a pKa of ≈4.8, which means that at neutral pH, the methacrylic acid groups in the lattice are almost completely deprotonated, making it an anionic polymer. PMAA can act as a polyelectrolyte with the ability to absorb and hold water. It was shown that PMAA used for a magnetic nanoparticles surface coating prevents nanoparticles aggregation and agglomeration [169–171].

Mexeriwattana and co-authors [172] investigated the interactions of blood serum proteins with poly (methacrylic acid) (PMAA)-coated magnetite nanoparticles obtained by the co-precipitation method [173]. PMAA-coated and bare magnetite nanoparticles were subjected to interaction with fetal calf serum (FCS). It has been shown that modification of the surface with poly (methacrylic acid) reduces the degree of coverage of the material by blood serum proteins compared to pure magnetite nanoparticles [172].

5.1.7. Immobilization of Proteins on Nanoparticles Coated with poly(*N*-isopropylacrylamide) (PNIPAM)

Poly(*N*-isopropylacrylamide) is a temperature-sensitive polymer. It is widely studied due to its water solubility and lower critical solution temperature (LCST) close to the physiological temperature value (about 36.5–37.5 °C) [174]. Poly(*N*-isopropylacrylamide) has a relatively simple structure based on a hydrophobic skeleton and strongly hydrophilic amide groups (-$CONH_2$) substituted with isopropyl moiety [175]. When heated in water above 32 °C, it undergoes a reversible phase transition with a lower critical solution temperature (LCST) from the swollen hydrated to a contracted dehydrated state, losing about 90% of its volume. It was notice that magnetic nanoparticles coated with poly(*N*-isopropylacrylamide) (PNIPAM) are characterized by thermal resistance and high magnetization as well as antibacterial properties [174,176].

Shamim et al. published a study on the effects of temperature, pH, and ionic strength on the adsorption and desorption of bovine serum albumin (BSA) on magnetic nanoparticles coated with a PNIPAM layer [177]. The influence of the incubation temperature on the amount of bound protein was also investigated. It was found that the amount of adsorbed protein was greater at higher temperatures (Figure 14). This behavior is attributed to the hydrophobic and hydrophilic properties of the nanomagnetic particles above and below the PNIPAM LCST, respectively. The effect of pH was also investigated, and it was observed that less protein was adsorbed at higher pH value. It was probably due to the electrostatic repulsion force between the protein molecules and the polymer shell covering the nanoparticles. The maximum amount of protein was adsorbed near the isoelectric point of BSA. The process of BSA desorption from the surface of nanoparticles was also investigated. The results showed that more protein was desorbed when adsorption was performed at the lower temperatures with the yield exceeded 80%.

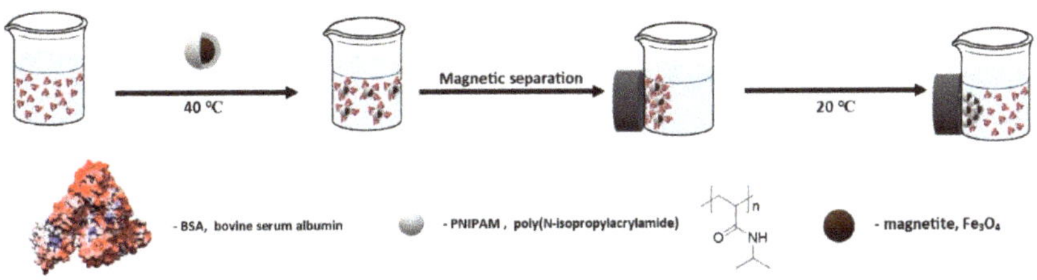

Figure 14. Adsorption and desorption scheme of protein on thermosensitive-polymer-coated magnetic particles.

An equally interesting potential application of poly(*N*-isopropylacrylamide) (PNIPAM)-coated magnetic nanoparticles has been described by Dionigi et al. [178]. The material obtained by co-precipitation was used as a matrix for the immobilization of vascular endothelial growth factor (VEGF). This article explores the possibility of delivering a cell growth factor, such as VEGF, under cell-friendly conditions to assure a high level of cell viability of HUVEC, (Primary Human Umbilical Vein Endothelial Cells) a well-known class of human cells. PNIPAM-coated magnetite nanoparticles incubated with HUVEC and loaded with VEGF demonstrated the release of the latter at 37–38 °C. The effect of the release of VEGF on the proliferation of cultivated HUVEC demonstrated both the loading and the preservation of the biological characteristics of the released VEGF. No adverse effect on cell proliferation was detected from the presence of MNPs. Summarizing, the PNIPAM-coated MNPs can be therefore considered as a promising material for controlled release of VEGF or other proteins able to stimulate vascular cells inside a scaffold.

5.2. *Immobilization of Proteins on Nanoparticles Coated with Natural Polymers*

Polymers of natural origin, the so-called biopolymers, are macromolecules especially used in biomedical sciences, including for the preparation of biocompatible and non-toxic materials. Due to the presence of reactive functional groups in their structure, these macromolecules can be easily subjected to chemical modification toward materials with better performance properties than the original biopolymer.

5.2.1. Immobilization of Proteins on Nanoparticles Coated with Chitosan (CS)

Chitosan (CS) is one of the most widely used biopolymers for magnetite core stabilization. This material is natural polysaccharide containing varying amounts of statistically decomposed structural units of 2-acetamido-2-deoxy-β-D-glucopyranose, (*N*-acetylglucosamine) and 2-amino-2-deoxy-β-D-glucopyranose ('-glucosamine) connected via β—(1→4)—glycosidic bonds [179]. Chitosan has proven antiviral, anti-inflammatory, analgesic, and antibacterial activity [180,181]. It is non-toxic, biocompatible, and biodegradable polymer. It was proven that coating magnetic nanoparticles with CS does not change the thermal and magnetic properties of the magnetite material. One-pot synthesis in the presence of low molecular weight CS showed that it can protect nanoparticles from aggregation due to electrostatic repulsion between positively charged nanoparticles. Chitosan-coated nanoparticles can easily penetrate cell membranes, which is often used in biomedical research. Free hydroxyl and amine groups of chitosan allow the surface modification of nanoparticles [182–185]. In an acidic environment, free CS amino groups gain a positive charge, thanks to which they can react with negatively charged groups of nucleic acids, which are used, among others, in MRI imaging [186]. Park et al. in the review on chitosan describe its numerous applications in the delivery of low molecular weight drugs [187].

There are many literature reports on the synthesis of chitosan-coated magnetic nanoparticles for the immobilization of catalytic proteins. Liang and Zhang [188] carried out the immobilization of papain from *Carica papaya* on magnetic nanoparticles coated with carboxymethyl chitosan. Magnetite nanoparticles were obtained by co-precipitation, and chitosan, before coating, was modified with monochloroacetic acid. Next, the chemical immobilization was performed and 1-ethyl-3-(3-dimethylaminopropyl) carbodiimide (EDC)/N-hydroxysulfosuccinimide (Sulfo-NHS) was used to couple the enzyme with the polymer (Figure 15). Conjugated papain showed increased enzymatic activity, better tolerance to pH and temperature changes, and increased storage stability compared to the native enzyme form.

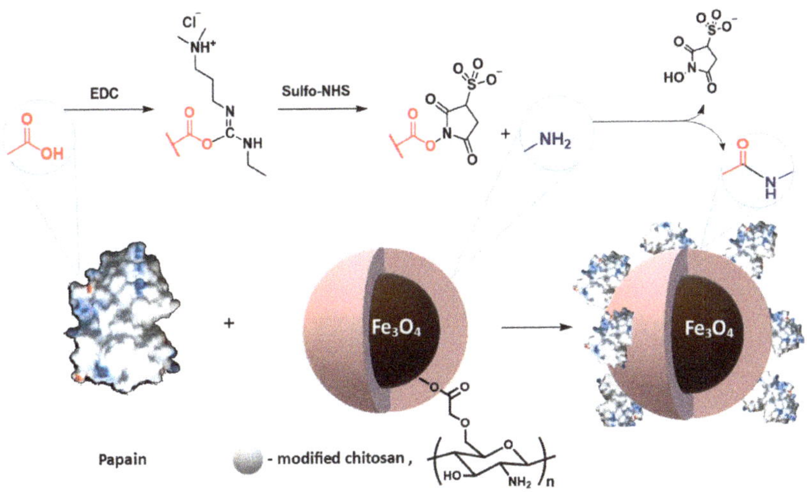

Figure 15. Schematic illustration for the surface modification of Fe_3O_4 magnetic nanoparticles by carboxymethylated chitosan and subsequent conjugation with papain.

In 2008, Li and co-authors synthesized chitosan-coated magnetic nanoparticles for the immobilization of *Saccharomyces cerevisiae* almond dehydrogenase (SCMD) [189]. Dehydrogenase was physically immobilized on the MNPs surface, and the effect of adsorption on its activity was examined. For this purpose, an SCMD-catalyzed reduction reaction of phenylglyoxylic acid to (R)-mandelic acid was used. After determination of the enzyme activity in the free and immobilized form at the temperature of 25 ° C and pH 7.0, it was shown that after immobilization, the enzyme retained about 50% of its activity compared to the free form of the protein. The possibility of immobilized SCMD reuse was also investigated. After each cycle, the immobilized SCMD was recovered by magnetic separation and recycled for the reduction of phenylglyoxylic acid. The activity of the first batch was taken as 100%. After seven catalytic cycles, the remaining activity was about 48.26% of the first use.

Wang [190] and Li and co-authors [189] obtained magnetic nanoparticles coated with chitosan cross-linked with glutaraldehyde and used them as a support for *Alcalase 2.4L* alkaline protease. The immobilization of *Alcalase 2.4L* alkaline protease on chitosan-coated magnetic nanoparticles caused an increase in enzyme activity, and the optimal range of temperature and pH profile was also significantly extended. As a model reaction to check enzyme activity, the hydrolysis of soy protein isolate (SPI) was performed by free and immobilized enzyme. The test results showed that the degree of SPI hydrolysis after 140 min was 18.38% for the immobilized enzyme and 17.50% for the free enzyme form. In addition, immobilized alkaline protease *Alcalase 2.4L* maintained approximately 86% of its original activity after ten cycles of reuse.

In turn, Sojitra et al. used chitosan-coated MNPS for pectinase immobilization [191]. Enzyme was chemically bonded to the MNPs surface with dextran polyaldehyde as a linker. Immobilization parameters such as linker concentration, time of immobilization, and support to enzyme ratio were optimized. Studies have shown that the thermal stability of pectinase immobilized on the surface of nanoparticles is twice as high as that of the free enzyme in the 55–75 °C temperature range. Moreover, the activity of immobilized pectinase was about 85% after seven reuse cycles and retained up to 89% after fifteen days of storage. Next, the obtained biocatalytic system was used for apple juice clarification. During this process, the turbidity was reduced to 74% after 150 min of juice treatment. In addition to obtaining a catalytic system that preserves the enzyme activity, this work showed that dextran polyaldehyde is a good linker for the chemical immobilization of enzymes on the surface of chitosan-coated nanoparticles. This immobilization technique includes a renewable and biocompatible natural biopolymer as a functionalizing and linking agent, making it an environmentally friendly technique and safer for workers compared to traditional chemicals.

Articles describing the immobilization of lipases on magnetic nanoparticles coated with chitosan have been published by Kuo, Wang, Monteiro, Hosseini, Ziegler-Borowska, Sikora, and Siodmiak [182,183,192–197].

Kuo et al. [195] used the response surface methodology (RSM) to find the optimal lipase immobilization conditions and to investigate the factors affecting the activity of the immobilized enzyme. It was shown that the optimal immobilization conditions were 2.14 h immobilization time, pH 6.37, and enzyme/carrier ratio 0.73 (*w*/*w*). The highest lipase activity was 20 U/g of chitosan-coated nanoparticles. Additionally, after twenty repeated cycles, the immobilized lipase retained over 83% of its initial activity. The immobilized enzyme showed better operational stability, including wider thermal and pH ranges than native protein, and it remains stable after 13 days of storage at 25 °C.

Wang et al. [196] describe the immobilization of lipase from *Thermomyces lanuginosus*. Chitosan-coated magnetic nanoparticles were prepared by a simple in situ co-precipitation that was used to covalently immobilize the enzyme via chemical conjugation after electrostatic entrapment (CCEE) (Figure 16).

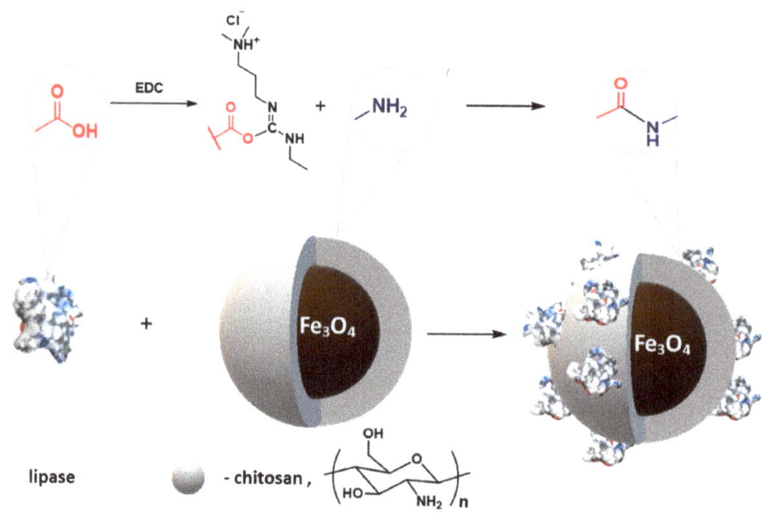

Figure 16. Scheme of lipase immobilization on the magnetic nanoparticles (MNPs) surface using 1-ethyl-3-(3-dimethylaminopropyl) carbodiimide (EDC)/sulfo-*N*-hydroxysuccinimide (NHS).

As optimal immobilization conditions, the protein/carrier ratio 19.8 mg/g, pH 5.0, time 4 h, and temperature 30 °C were used. A high immobilization efficiency at the level of 75% and a bounded protein amount of 16.8 mg/g of the carrier were obtained. Moreover, the immobilized lipase retained about 70% of its initial activity after ten catalytic cycles. The prepared catalytic system was used for the synthesis of ascorbyl palmitate, which resulted with above 50% conversion of ascorbic acid to the appropriate ester. Based on these data, it was noticed that the immobilization of lipase on magnetic nanoparticles coated with chitosan by the CCEE method is an efficient and simple way to obtain a stable catalytic system.

Monteiro et al. [197] investigated the immobilization of lipase A from *Candida antarctica*. Chitosan-coated magnetite nanoparticles were activated with glutaraldehyde, and Lipase A from *Candida antarctica* was attached to the carrier surface by covalent bonding with 84%

immobilization efficiency. Additionally, the immobilized biocatalyst showed a half-life about 8–11 times longer than for the free enzyme at solution with pH 5–9 and greater activity at almost all tested pH values.

Hosseini et al. [194] and Monteiro [197] synthesized chitosan coated magnetic nanoparticles in two steps. In the first step, $FeCl_3 \cdot 6H_2O$ and sodium acetate reacted in ethylene glycol, and then the polymer shell was crosslinked with citric acid with the use hydroxyl and amine groups of chitosan at neutral pH. The resulting nanoparticles were used for Lipase B from *Candida antarctica* immobilization with glutaraldehyde as the linker (Figure 17). The results showed that the immobilized enzyme has higher storage stability than free protein. It has also been successfully used to itaconic anhydride oligomerization by ring-opening esterification. This process is a green approach for the production of functional oligoesters and can be applied to make photo-curable esters.

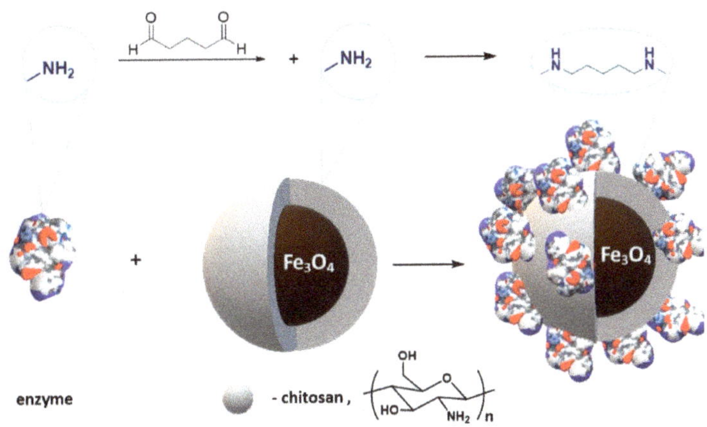

Figure 17. Enzyme immobilization on chitosan-coated magnetite nanoparticles with glutaraldehyde as a linker.

Ziegler-Borowska et al. [182] synthesized magnetite nanoparticles with surface modified with a mixture of two polymers: chitosan and poly [N-benzyl-2-(methacryloxy)-N, N-dimethylethanaminium bromide] (PQ) (Figure 18).

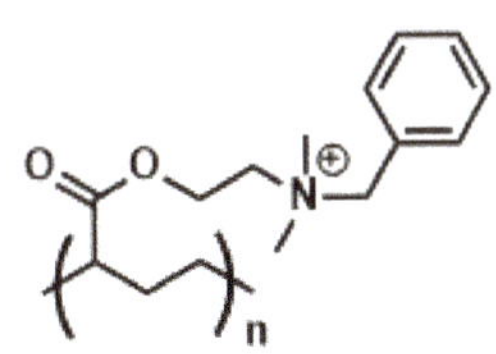

PQ

Figure 18. Structure of poly [N-benzyl-2-(methacryloxy)-N, N-dimethylethanaminium bromide] (PQ).

Nanoparticles with different mass ratios of these polymers were used for the immobilization of lipase from *Candida rugosa*. The enzyme was covalently bounded to the nanoparticle surface by EDC/sulfo-NHS activation. The activity recovery of the immobilized enzyme was estimated by the hydrolysis of olive oil, and its maximum value was about 82% for CS–PQ nanoparticles (1:1). The residual activity of immobilized lipase was over 90% after five catalytic cycles and remained at the level of 70–72.1% after 10 cycles. Moreover, it was noticed that the presence of the quaternary ammonium salt had a positive effect on the dispersion and stability of the solutions due to the spatial effects and electrostatic repulsion of polymer chains. As a result of quaternary ammonium salt biological activity, the studied magnetic nanoparticles may be also of particular importance in antimicrobial applications. On the other hand, Siódmiak and co-workers used this catalytic system for kinetic separation through enantioselective esterification of (R, S)-ibuprofen—a widely used non-steroidal anti-inflammatory drug [183]. Lipase, immobilized on the surface of CS/PQ magnetic carriers with (EDC)/(sulfo-NHS) activation showed high catalytic activity, which allowed obtaining (S)-methyl ibuprofen ester with high enantioselectivity (E = 50.6). Moreover, the properties of this magnetite particles allow for better optimization of the enantioselective esterification of (R, S)-ibuprofen and as a result could reduce the overall cost of this reaction. In addition, the use of this nanocatalytic system allowed for maintaining high enantioselective activity after repeated use.

Later, Ziegler-Borowska and co-authors prepared magnetite nanoparticles coated with chemically-modified chitosan rich in free amino groups remote from the polymer chain (Figure 19) [183,185,198–202]. First, chitosan-coated magnetite nanoparticles were synthesized by an in situ co-precipitation reaction in an alkaline solution. Then, the reactive chitosan groups, hydroxyl and amino, were used for its functionalization, obtaining three polymer coatings with different content of amino groups able to bind protein (Figure 19). Prepared nanoparticles were used for the chemical immobilization of proteins: lipase from *Candida rugosa* [183,193,201], human serum albumin (HSA) [185,199], and the androgen receptor (AR) [202].

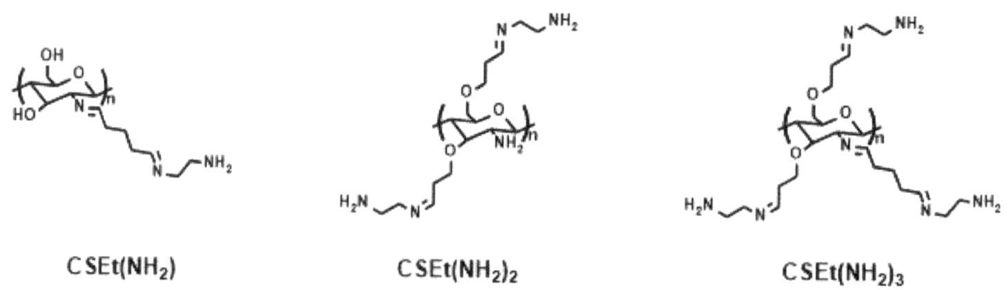

Figure 19. Structure of the aminated chitosan Fe_3O_4-CSEt$(NH_2)_{1-3}$.

Sikora and co-authors [193,201] used these magnetic nanoparticles with immobilized lipase from *Candida rugosa* for the enantioselective acetylation of (R, S)-atenolol. Additionally, the catalytic activity of two types of commercially available lipases from *Candida rugosa* immobilized on two different magnetic nanoparticles was compared. Among all the tested catalytic systems, the best results were obtained with the lipase from *Candida rugosa* immobilized on Fe_3O_4-CS-Et$(NH_2)_1$ (E = 66.9, c = 41.84%, ee = 94.1%). Additionally, these studies also showed that even after five catalytic cycles, the immobilized lipase maintains high catalytic activity. Next, Marszałł et al. [202] used these same supports for immobilization of the androgen receptor (AR). The research involved comparing various AR carriers, such as silica-coated magnetic nanoparticles and chitosan-coated nanoparticles with varying amounts of amino groups. Immobilization was performed in two ways: by covalent immobilization of the AR via an amino terminal group or available carboxyl groups. The initial characterization of AR-coated magnetic nanoparticles was performed with dihydrotestosterone as a well-known AR ligand. Subsequently, chitosan-modified nanoparticles with distant primary amine groups (Fe_3O_4-CS-Et$(NH_2)_3$) were used for the isolation of AR ligands (bicalutamide, flutamide, hydroxyflutamide, and levonogestrel) from the mixture. Based on the obtained results, it was noticed that the selected nanoparticles are a promising semi-quantitative tool for the identification of compounds with high affinity for AR and may be of particular importance in the identification of new agonists or antiandrogens. Finally, these materials were used for HSA immobilization. Moreover, the influence of the selected method of functionalization of the polymer coating covering magnetite nanoparticles (MNP) as well as drying methods of the carrier on the immobilization of human serum albumin HSA was performed [185]. Albumin was immobilized on three types of nanoparticles coated with aminated chitosan with different content of amino groups at a large distance from the Fe_3O_4-CS-Et$(NH_2)_{1-3}$ surface. It was noticed that both the synthesis method and the method of drying nanoparticles have a large impact on the effectiveness of immobilization. The largest amount of protein was immobilized on Fe_3O_4-CS-Et $(NH_2)_3$, and it was 210.32 mg/g nanoparticles. In the case of the materials Fe_3O_4-CSEt(NH_2) and Fe_3O_4-CS-Et$(NH_2)_3$ dried by the freeze-drying method, the efficiency of protein immobilization was 200% higher than in the case of nanoparticles dried in a vacuum dryer. In another work, Ziegler-Borowska et al. used such immobilized albumin for protein–drug interaction study [184].

The synthesis of magnetite nanoparticles coated with a mixture of chitosan and collagen with the use of various cross-linking agents of the polymer coating, glutaraldehyde (Glu) and squaric acid (SqA), was also described by Ziegler-Borowska and co-authors [192]. Obtained nanomaterials were prepared in a standard co-precipitation reaction and then coated with chitosan (CS), collagen (Coll), and a mixture of these two biopolymers and cross-linked. As a result of the performed syntheses, six types of nanomaterials were obtained: CS/Glu (Fe_3O_4), Coll/Glu (Fe_3O_4), CS–Coll/Glu (Fe_3O_4), Coll/SqA(Fe_3O_4), CS/SqA (Fe_3O_4), and CS–Coll/SqA(Fe_3O_4). Subsequently, lipase from *Candida rugosa* immobilization was carried out on the surface of the obtained nanoparticles with EDC/NHS activation. The best results in terms of recovered lipase activity and specific activities were observed for nanoparticles with a polymer coating cross-linked with squaric acid. The specific activity of lipase immobilized on SqA cross-linked materials was 52 U/mg lipase, and it was about two times higher than for the enzyme immobilized on nanomaterial with glutaraldehyde 26 U/mg lipase. Additionally, after the fifth catalytic cycle, the residual activity for all tested magnetic nanoparticles was about 80–90%, and at the end of the tenth cycle, the immobilized lipases still retained almost 80% of their activity. Moreover, a little hyperactivation of lipase immobilized on CS/SqA (Fe_3O_4) and CS–Coll/SqA(Fe_3O_4) nanoparticles was observed.

5.2.2. Immobilization of Proteins on Nanoparticles Coated with Agarose

Agarose is a polysaccharide usually obtained from some red seaweed. It is a linear polymer composed of the agarobiose repeating unit, which is a disaccharide consisting of D-galactose that is well soluble in water 3,6-anhydro-L-galactopyranose (Figure 20) [203,204].

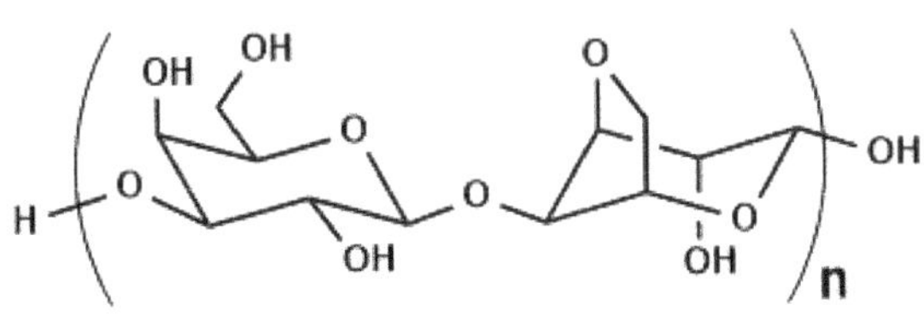

Figure 20. The structure of agarose.

After cooling of its aqueous solution, this inert polymer forms thermo-reversible gels. The biocompatibility of agarose hydrogels and the mild conditions required for their gelling make the composite hydrogel widely used in the biomedical industry. As a result of the non-toxicity of agarose-coated MNPs, they are widely used in drug delivery systems [205–207].

Chen and co-authors synthesized agarose-coated magnetic nanoparticles for β-glucosidase (BGL) immobilization [208]. Magnetite nanoparticles coupled with agarose (AMNPs) were prepared via co-precipitation reaction in alkaline condition with span–80 surfactants addition in organic solvent. Next, iminodiacetate was attached to the MNPs through epichlorohydrin agent and then chelated with metal ions. Enzyme immobilization was performed by the physical adsorption. Cobalt ion chelated nanoparticles prepared by this way showed a high enzyme adsorption capacity of 1.81 mg/g nanoparticles. Moreover, a hyperactivation of immobilized β-glucosidase was observed (enzyme activity of 117% per gram of protein in β-glucosidase immobilization). Additionally, compared to free BGL, the immobilized enzyme showed higher hydrolytic activity and thermostability and better operational stability. After running for 15 catalytic cycles, this catalytic system retained over 90% of its initial activity.

5.2.3. Immobilization of Proteins on Nanoparticles Coated with Starch

Starch is one of the most interesting natural polysaccharides used for MNPs coating. It is a plant polysaccharide, consisting only of glucose units linked by α-glycosidic bonds, acting as an energy store in plants [209]. Starch hydrolyses only to α-D-glucose but is not a chemically homogeneous compound—it consists of two fractions: unbranched amylose and branched amylopectin (Figure 21).

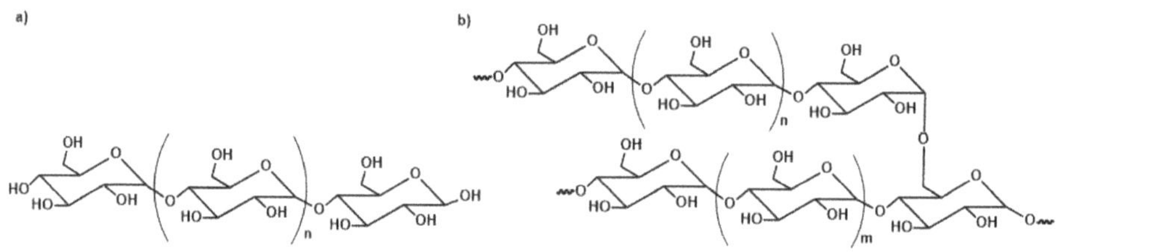

Figure 21. Chemical structure of starch components: (**a**) amylopectin and (**b**) amylose.

The presence of highly reactive hydroxyl groups and the relatively good solubility of starch make it susceptible to modifications, which inspires the creation of new materials with potential properties such as biodegradability and non-toxicity [210]. For chemical modifications of this polysaccharide oxidation, esterification, and etherification reactions are used [211,212]. It was also confirmed that starch-coated magnetite nanoparticles retain their magnetic properties [213] and are characterized with high biocompatibility as well as colloidal stability. As a result of that, starch and modified starch-coated magnetite nanoparticles can be considered as one of the biomaterials with biological potential for applications in the biomedical department [214,215].

Ziegler-Borowska [212] carried out the synthesis of starch and modified starch-coated magnetite nanoparticles. First, starch-coated MNPs were obtained by in situ co-precipitation, and the polymer shell was modified in the next stage: it was oxidized to dialdehyde starch and then simply and quickly aminated without solvent (pounding in a mortar) to nanoparticles coated with starch enriched with amino groups (Figure 22).

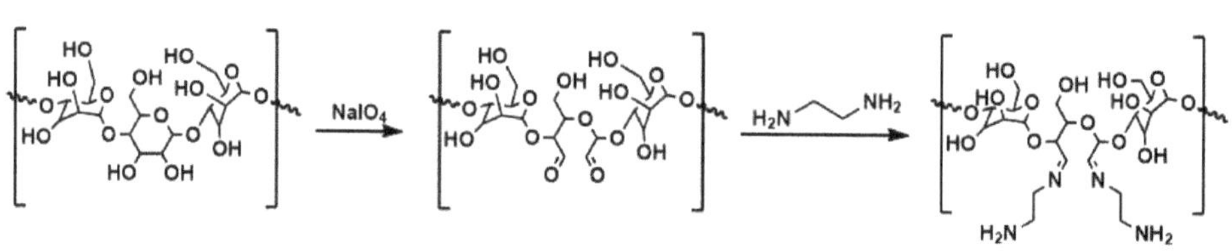

Figure 22. Synthesis of aminated starch.

These magnetite nanoparticles were used as a carrier for HSA immobilization with glutaraldehyde as the binding agent. The amount of HSA immobilized on the nanoparticles was about 149.96 mg HSA/g of the carrier, and the protein retained its activity after immobilization. Next, these HSA-aminated starch-coated MNPs were used for HSA-ketoprofen binding study [185].

Wang et al. used starch-coated magnetic nanoparticles for pectinase produced by Penicillium oxalicum F67 (PoPase) immobilization [216]. Before the immobilization process, magnetic nanoparticles coated with corn starch were cross-linked with glutaraldehyde. Enzyme immobilization on the support was performed by physical adsorption. Additionally, the influence of various factors on the rate of enzyme regeneration after immobilization was investigated. Next, free and immobilized enzymes were used to extract apple juice to evaluate their effect on juice yield. The studies showed that after eight catalytic cycles, the immobilized enzymes retained about 60% of the initial PoPase activity. This confirmed that starch-coated magnetic nanoparticles are a good carrier for PoPase immobilization, which may be applicable to juice processing.

Gagnon and co-authors [217] described the use of starch-coated magnetic nanoparticles for the capture of monoclonal IgG antibodies. A hybrid system that embodies elements of both PEG precipitation and steric exclusion chromatography on starch-coated 200 nm nanoparticles enables the magnetic capture of 78 mg IgG per mg of particles, with 98% antibody recovery.

Wensheng Lu and co-authors published the synthesis of magnetic nanoparticles coated with dialdehyde starch [218]. They oxidized starch with sodium periodate; then, the nanoparticles were cross-linked with epichlorohydrin solution, and bovine serum albumin (BSA) was successfully immobilized on the carrier surface. The support prepared in this way can be used for targeted drug release.

In 2008, Kuroiwa and co-authors described the immobilization of the enzyme chitosanase on magnetic nanoparticles coated with amylose [219]. As previously mentioned, amylose is a polysaccharide that is a constituent of starch. Enzyme immobilization on the support was carried out by two different techniques—physical adsorption and multiple covalent bonding at different concentrations of the immobilized enzyme. The yield of the chemical immobilization was higher than that of physical adsorption. A high protein immobilization capacity (maximum 0.4 g protein/g molecule) was observed for these nanoparticles. Next, chitosanase immobilized on magnetic amylose-coated nanoparticles was used to produce chitosan oligosaccharides with high physiological activity. The physiologically active pentamers and hexamers of chitosan oligosaccharide were obtained by this way in high yield (40%), which was higher than for conventional methods.

5.2.4. Immobilization of Proteins on Nanoparticles Coated with Cellulose

Cellulose is an unbranched biopolymer composed of D-glucose units linked linearly with β-1,4-glycosidic bonds [220]. Cellulose is the main structural component of plants and can also be produced by marine animals, algae, fungi, bacteria, invertebrates, and the pulpit, which make it the most abundant polymer in nature [221,222]. Coating nanoparticles with cellulose results in the formation of a hydrophilic surface that prevents agglomeration, increases the dispersion of nanoparticles in physiological solution, and increases bioavailability; it is also a suitable coating for functionalization [223].

Aguilera and co-authors showed that magnetic nanoparticles coated with carboxymethyl cellulose—a cellulose derivative—have great potential as drug delivery systems during neurological treatment [223]. The biodegradability of the coating allows the drug to be delivered at a controlled and steady rate to the target site. Moreover, such materials are able to cross the blood–brain barrier (BBB) [223].

Namdeo and co-authors described the synthesis of magnetic nanoparticles coated with cellulose for the immobilization of α-amylase [224]. Before immobilization, the cellulose coating was subjected to chemical modifications-oxidation with periodic acid (Figure 23).

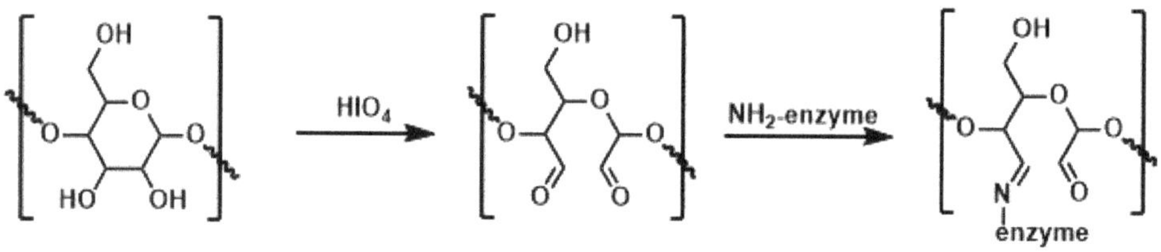

Figure 23. Scheme of dialdehyde cellulose synthesis and enzyme immobilization.

Amylase was covalently immobilized on this support, and it was noticed that nanoparticles coated with 16 and 28 wt % of cellulose bounded about 9.1 and 16.2 mg enzyme/g nanoparticles, respectively. Therefore, it has been proven that immobilized α-amylose retains its ability for starch degradation. Later, Ivanova and co-authors also published the results of immobilization of α-amylase on magnetic nanoparticles coated with cellulose [225]. Another application of these magnetic nanoparticles was presented by Anirudhan and co-authors [226]. Magnetic nanoparticles coated with cellulose were used for heme protein adsorption. Nanoparticles were obtained by co-precipitation with nanocrystalline cellulose, which was further modified by free radical graft copolymerization using VSA (vinylsulfonic acid) and MAA (methacrylic acid). It has been found that this material can be an effective myoglobin adsorbent from aqueous solutions. The adsorption was affected by parameters such as time, temperature, concentration, and pH. The best recovery of myoglobin was achieved at pH 7.0, near the isoelectric point of the protein, indicating the existence of electrostatic interaction between the adsorbent and the protein.

Guo and co-authors used magnetic nanoparticles coated with modified cellulose for protein separation [227]. First, primary hydroxyl groups on the cellulose surface were selectively oxidized with TEMPO (2,2,6,6-tetramethylpiperidinyloxy). At the same time, magnetite nanoparticles functionalized with amino groups were synthesized by co-precipitation reaction and coupled to previously modified cellulose via EDC/sulfoNHS linking. In the next step, MNPs were complexed with copper ions to provide specific binding sites for proteins. The prepared material showed high lysozyme binding capacity (860.6 ± 14.6 mg/g) as well as efficient reutilization for protein separation. Total protein recovery using simple elution was approximately 98%.

Lately, Mohammadi and co-authors focused on the interaction of bovine α-lactalbumin (BLA) with variously coated magnetic nanoparticles: naked, dopamine stabilized, and coated with cellulose [228]. They showed that magnetic nanoparticles coated with cellulose have the highest protein binding affinity compared to the rest. Additionally, the circular dichroism spectra revealed that the BLA conformation was preserved after interaction with magnetic nanoparticles. This important feature, in addition to the exceptionally strong binding affinity, makes these particles novel model nanostructures for nanomedicine-focused applications.

5.2.5. Immobilization of Proteins on Nanoparticles Coated with Dextran

Dextran is another natural polysaccharide-based polymer applied for magnetite nanoparticles coating [229–231]. It is a water-soluble macromolecule consisting of many glucose molecules linked by α-1,6-glycosidic bonds. It was shown that magnetic nanoparticles coating with hydrophilic organic polymers such as dextran improve their drug delivery properties and colloidal stability [232–234]. Moreover, hydroxyl groups of dextran structure can be modified and functionalized with primary amines, which affects their ability to bind targeted ligands [235].

Weissleder and al. showed that magnetic nanoparticles coated with transferrin (Tf) could be used to image the transferrin receptor by magnetic resonance imaging (MRI) [236]. These results can be related to the expression of this receptor. After the administration of magnetic nanoparticles conjugated with transferrin to the body, their high accumulation in cancer cells was observed, which allows their use as a probe for imaging neoplastic sites by MRI [236]. Despite high doses of the probe, the observed signal in MRI imaging showed very low values. Therefore, in the same year, Hogemann et al. describe the synthesis of novel nanoparticles for magnetic resonance imaging (MRI) of transferrin receptors characterized by a much better signal [237]. These nanoparticles were based on magnetite core coated with dextran and immobilized transferrin. The best imaging results were obtained for MNPs coated with cross-linked dextran (CL10) and immobilized oxidized transferrin (Tf-SH) with N-succinimidyl 3-(2-pyridyldithio) propionate (SPDP) as a linker. It was showed that the use of this linker does not affect the activity of immobilized transferrin (Tf). It was also proved that in such a system, four transferrin (Tf) molecules were immobilized on one nanoparticle surface. This make it possible to reduce the amount of potential dose injected into the body in order to obtain the appropriate intensity of imaging [237].

An equally interesting use of magnetic nanoparticles coated with dextran was presented by Horng et al., which used this system for the anti-C-reactive protein (CRP) antibody immobilization [238]. Prepared magnetic nanoparticles were applied to the C-reactive protein (CRP) detection with use a SQUID (superconducting quantum interference device) gradiometer. It is well known that C-reactive protein (CRP) is produced by the body during tissue damage, necrosis, and ongoing inflammation, and it is a non-specific marker of these phenomena. Standard for CRP detection, commercially available ELISA [239] assays are used. Results obtained for dextran-coated MPNs showed CRP detection at about 10 ng/mL. This value, compared to standard procedures using ELISA tests, is lower by an order of magnitude, which significantly improves the measurement possibilities [238].

The immobilization of the antibody on carboxymethylated dextran (CMD)-coated magnetic nanoparticles was presented by Li et al. (Figure 24) [240]. Anti-BSA was immobilized on the nanoparticles surface with EDC/NHS activation to couple the amino groups of the antibody and CMC carboxyl groups (Figure 24). Next, this system was tested for the ability to trap BSA.

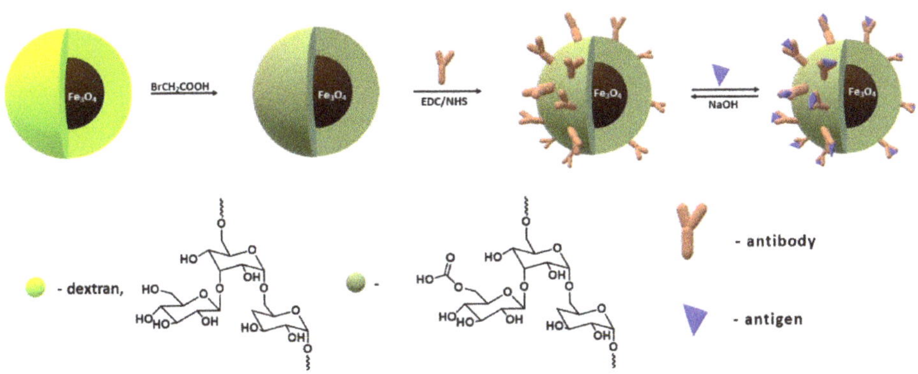

Figure 24. Synthesis of carboxymethylated dextran-coated magnetic nanoparticles (CMD-MNPs) covered with anti-bovine serum albumin (BSA) antibody and subsequent capture and release of BSA in sample solution.

The results obtained in this study indicate that over 20 mg of BSA was bounded per 1 g of anti-BSA immobilized on CMD-coated nanoparticles. Moreover, the potential use of such systems for selective proteins capture was mentioned [240].

Carboxymethylated dextran (CMD) was also used by Vasić and co-authors in their research for alcohol dehydrogenase (ADH) immobilization [234]. It was shown that the remaining ADH activity after immobilization at 4 °C oscillates around only 25%.

Ziv and co-authors [241] describe the body's reactions measured by the amount of antibodies produced against magnetic nanoparticles coated with dextran and gelatin with immobilized HSA (human serum albumin) (Gel-MNP-Dex-HSA). The synthesis of such a system was performed with the innovative method of nucleation. This procedure consists of a controlled growth of magnetic nanoparticles coated with gelatin from a gelatin solution to which iron salts II and III have been added. In the final stage, HSA was immobilized using the previously prepared coating and the Michael addition reaction. The body's response to these systems was tested by measuring the number of produced antibodies (anti-gelatin, anti-dextran, and anti-HSA). In conclusion, the authors indicate the correctness of using the gelatin nucleation method due to the very weak immunogenic nature of this polymer. However, regarding dextran, it was found that it could be replaced with another biopolymer of lower immunogenicity. Investigating the immunogenicity of such materials is extremely important from the point of view of biomedical applications; therefore, Ziv et al. plan further studies of nanomaterials in this area [241].

6. Conclusions

Summarizing the collected information, it can be stated that every year, the use of magnetite nanoparticles coated with polymers for protein immobilization is increasing. The design, synthesis, and application of nanoparticles coated with biopolymers in their original form and after chemical modifications are developing particularly dynamically. Among the proteins, enzymes are still most often immobilized on magnetite particles, which is most likely due to the possibility of their reuse in the catalytic cycle. Nevertheless, there are more and more works on the immobilization of serum proteins and antibodies. Table 4 below summarizes all the most important applications of magnetite nanoparticles in protein immobilization discussed in the above work. Such a summary, divided into the polymer covering the magnetite core and the immobilized protein, can be of convenience for the reader.

Table 4. List of proteins immobilized on magnetic nanoparticles coated with polymers described in the literature.

Protein	Polimer	References
Fetal bovine serum (FBS)	polyethylene glycol (PEG)	[125]
Cytochrome C	polyethylene glycol (PEG)	[124]
Transferrin	polyvinyl alcohol (PVA)	[131]
Horseradish peroxidase	polyvinyl alcohol/chitosan (PVA/CS)	[132]
Trypsin	polyvinyl alcohol (PVA)	[134]
Herceptin antibody (HER)	poly(D,L-lactide-co-glycolide) (PLGA)/polyvinyl alcohol (PVA)	[143]
Bovine serum albumin (BSA)	poly(D,L-lactide-co-glycolide)(PLGA)	[145]
Laccase from *Trametes versicolor*	polyethyleneimine (PEI)	[149]
Green fluorescent protein	polyethyleneimine (PEI)	[154]
Fetal bovine serum (FBS)	polyethyleneimine (PEI)	[155]
Candida rugosa lipase i *Mucor miehei* lipase	polyethyleneimine (PEI)	[156]
Fetal bovine serum (FBS)	polyethyleneimine (PEI)	[158]
Fetal bovine serum (FBS)	polyacrylic acid (PAA)	[158]
Fetal bovine serum (FBS)	polyacrylic acid (PAA)	[155]
Bovine serum albumin (BSA) and Immunoglobulin G	polyacrylic acid (PAA)/chitosan (CS)	[163]
Lipase from Candida Rugosa	polyacrylic acid (PAA)	[164]
Tissue plasminogen activator (rtPA)	polyacrylic acid (PAA)	[166]
Lysozyme (LYZ) and bovine serum albumin (BSA)	polyacrylic acid (PAA)	[167]
Fetal bovine serum (FBS)	polymethacrylic acid (PMAA)	[172]
Bovine serum albumin (BSA)	poly(*N*-isopropylacrylamide) (PNIPAM)	[177]
Vascular endothelial growth factor (VEGF)	poly(*N*-isopropylacrylamide) (PNIPAM)	[178]
Streptavidin	poly(*N*-isopropylacrylamide) (PNIPAM)	[242]
Papain from Carica papaya	Chitosan (CS)	[188]
Saccharomyces cerevisiae Mandelated dehydrogenase (SCMD)	Chitosan (CS)	[189]
Pectinase	Chitosan (CS)	[191]
Lipase from Candida rugosa	Chitosan (CS)	[195]
Lipase from *Thermomyces lanuginosus*	Chitosan (CS)	[196]
Lipase A from Candida antarctica	Chitosan (CS)	[197]

Table 4. *Cont.*

Protein	Polimer	References
Lipase B from Candida antarctica	Chitosan (CS)	[194]
Lipase from Candida rugosa	Chitosan (CS)	[182]
Lipase from Candida rugosa	Chitosan (CS)	[183]
Lipase from Candida rugosa and Human serum albumin (HSA)	Chitosan (CS)	[198]
Lipase from Candida rugosa	Chitosan (CS)	[193]
Lipase from Candida rugosa	Chitosan (CS)	[201]
Lipase from Candida rugosa	Chitosan (CS)	[192]
Androgen receptor (AR)	Chitosan (CS)	[202]
Human serum albumin (HSA)	Chitosan (CS)	[184]
Human serum albumin (HSA)	Chitosan (CS)	[185]
β-Glucosidase (BGL)	Agarose	[208]
Human serum albumin (HSA)	Starch	[212]
Pectinase	Starch	[216]
Immunoglobulin G	Starch	[217]
Bovine serum albumin (BSA)	Starch	[218]
Chitosanase	Amylose	[219]
α-amylase	Cellulose	[224]
α-amylase	Cellulose	[225]
Heme	Cellulose	[226]
Lysozyme	Cellulose	[227]
Bovine α-lactalbumin (BLA)	Cellulose	[228]
Transferrin	Dextran	[237]
Anti-CRP antibodies	Dextran	[238]
Anti-BSA	Dextran	[240]
Alcohol dehydrogenase (ADH)	Dextran	[234]
Human serum albumin (HSA)	Dextran	[241]

Author Contributions: Conceptualization K.M. and M.Z.-B.; data curation, K.M., P.N., M.Z.-B.; writing—original draft preparation, K.M., P.N., M.Z.-B.; writing—review and editing M.Z.-B.; visualization, P.R.; supervision, M.Z.-B.; project administration and funding acquisition, M.Z.-B. All authors have read and agreed to the published version of the manuscript.

Funding: This work was supported by the National Science Centre grant 2014/15/D/NZ7/01805.

Acknowledgments: K.M. P.N. and M.Z.-B. are members of Center of Excellence "Towards Personalized Medicine" operating under Excellence Initiative—Research University and are grateful to this Center for financing the costs of publication.

Conflicts of Interest: The authors declare no conflict of interest.

References

1. Jeevanandam, J.; Barhoum, A.; Chan, Y.S.; Dufresne, A.; Danquah, M.K. Review on nanoparticles and nanostructured materials: History, sources, toxicity and regulations. *Beilstein J. Nanotechnol.* **2018**, *9*, 1050–1074. [CrossRef] [PubMed]
2. Aseri, A.; Garg, S.K.; Nayak, A.; Trivedi, S.K.; Ahsan, J. Magnetic nanoparticles: Magnetic nano-technology using biomedical applications and future prospects. *Int. J. Pharm. Sci. Rev. Res.* **2015**, *31*, 119–131.
3. Dzhardimalieva, G.I.; Pomogailo, A.D.; Rozenberg, A.S.; Leonowicz, M. Magnetic Metallopolymer Nanocomposites: Preparation and Properties. *Magn. Nanoparticles* **2009**, 59–85. [CrossRef]
4. Kolesnichenko, V.L. *Synthesis of Nanoparticulate Magnetic Materials*; Wiley-VCH: Weinheim, Germany, 2009; ISBN 9783527407903.
5. Vedmedenko, E. *Competing Interactions and Patterns in Nanoworld. The Chemistry of Nanomaterials Nanoparticles Introduction to Nanotechnology*; Wiley-VCH: Weinheim, Germany, 2007; ISBN 9783527404841.
6. Can, M.M.; Coşkun, M.; Firat, T. A comparative study of nanosized iron oxide particles: Magnetite (Fe_3O_4), maghemite (γ-Fe_2O_3) and hematite (α-Fe_2O_3), using ferromagnetic resonance. *J. Alloys Compd.* **2012**, *542*, 241–247. [CrossRef]
7. Mahdavi, M.; Ahmad, M.B.; Haron, M.J.; Namvar, F.; Nadi, B.; Ab Rahman, M.Z.; Amin, J. Synthesis, surface modification and characterisation of biocompatible magnetic iron oxide nanoparticles for biomedical applications. *Molecules* **2013**, *18*, 7533–7548. [CrossRef]

8. Zhang, Z.; Kong, J. Novel magnetic Fe_3O_4@C nanoparticles as adsorbents for removal of organic dyes from aqueous solution. *J. Hazard. Mater.* **2011**, *193*, 325–329. [CrossRef]
9. Ito, A.; Shinkai, M.; Honda, H.; Kobayashi, T. Medical application of functionalized magnetic nanoparticles. *J. Biosci. Bioeng.* **2005**, *100*, 1–11. [CrossRef]
10. Garcia-Galan, C.; Berenguer-Murcia, Á.; Fernandez-Lafuente, R.; Rodrigues, R.C. Potential of different enzyme immobilization strategies to improve enzyme performance. *Adv. Synth. Catal.* **2011**, *353*, 2885–2904. [CrossRef]
11. Liu, X.; Zhang, L.; Zeng, J.; Gao, Y.; Tang, Z. Superparamagnetic nano-immunobeads toward food safety insurance. *J. Nanoparticle Res.* **2013**, *15*. [CrossRef]
12. Rahim, S.; Iftikhar, F.J.; Malik, M.I. *Biomedical Applications of Magnetic Nanoparticles*; Elsevier: Amsterdam, The Netherlands, 2020; pp. 301–328. ISBN 9780128169605.
13. Magro, M.; Venerando, A.; Macone, A.; Canettieri, G.; Agostinelli, E.; Vianello, F. Nanotechnology-based strategies to develop new anticancer therapies. *Biomolecules* **2020**, *10*, 735. [CrossRef]
14. Long, N.V.; Yang, Y.; Teranishi, T.; Thi, C.M.; Cao, Y.; Nogami, M. Biomedical applications of advanced multifunctional magnetic nanoparticles. *J. Nanosci. Nanotechnol.* **2015**, *15*, 10091–10107. [CrossRef]
15. Banerjee, R.; Katsenovich, Y.; Lagos, L.; McIintosh, M.; Zhang, X.; Li, C.-Z. Nanomedicine: Magnetic Nanoparticles and their Biomedical Applications. *Curr. Med. Chem.* **2010**, *17*, 3120–3141. [CrossRef] [PubMed]
16. Wallyn, J.; Anton, N.; Vandamme, T.F. Synthesis, principles, and properties of magnetite nanoparticles for in vivo imaging applications—A review. *Pharmaceutics* **2019**, *11*, 601. [CrossRef]
17. Rossi, L.M.; Costa, N.J.S.; Silva, F.P.; Wojcieszak, R. Magnetic nanomaterials in catalysis: Advanced catalysts for magnetic separation and beyond. *Green Chem.* **2014**, *16*, 2906–2933. [CrossRef]
18. Wu, W.; He, Q.; Jiang, C. Magnetic iron oxide nanoparticles: Synthesis and surface functionalization strategies. *Nanoscale Res. Lett.* **2008**, *3*, 397–415. [CrossRef] [PubMed]
19. Ali, A.; Zafar, H.; Zia, M.; ul Haq, I.; Phull, A.R.; Ali, J.S.; Hussain, A. Synthesis, characterization, applications, and challenges of iron oxide nanoparticles. *Nanotechnol. Sci. Appl.* **2016**, *9*, 49–67. [CrossRef] [PubMed]
20. Xu, J.; Sun, J.; Wang, Y.; Sheng, J.; Wang, F.; Sun, M. Application of iron magnetic nanoparticles in protein immobilization. *Molecules* **2014**, *19*, 11465–11486. [CrossRef] [PubMed]
21. Shabatina, T.I.; Vernaya, O.I.; Shabatin, V.P.; Melnikov, M.Y. Magnetic Nanoparticles for Biomedical Purposes: Modern Trends and Prospects. *Magnetochemistry* **2020**, *6*, 30. [CrossRef]
22. Rehse, K. Kurzmi tteilung Vergleichende Untersuchungen zur Bindung nichtsteroidaler Antirheumatika an Humanserumalbumin und deren Interaktion rnit Phenprocoumon. *Archiv der Pharmazie* **1989**, *7*, 241–243.
23. Holm, J.; Babol, L.N.; Markova, N.; Lawaetz, A.J.; Hansen, S.I. The interrelationship between ligand binding and thermal unfolding of the folate binding protein. the role of self-association and pH. *Biochim. Biophys. Acta Proteins Proteom.* **2014**, *1844*, 512–519. [CrossRef]
24. Wahab, R.A.; Elias, N.; Abdullah, F.; Ghoshal, S.K. On the taught new tricks of enzymes immobilization: An all-inclusive overview. *React. Funct. Polym.* **2020**, *152*, 104613. [CrossRef]
25. Rehm, F.B.H.; Chen, S.; Rehm, B.H.A. Enzyme engineering for in situ immobilization. *Molecules* **2016**, *21*, 1370. [CrossRef] [PubMed]
26. Hoarau, M.; Badieyan, S.; Marsh, E.N.G. Immobilized enzymes: Understanding enzyme-surface interactions at the molecular level. *Org. Biomol. Chem.* **2017**, *15*, 9539–9551. [CrossRef] [PubMed]
27. Leitgeb, M.; Knez, Ž.; Vasić, K. Micro-and Nanocarriers for Immobilization of Enzymes. In *Micro and Nanotechnologies for Biotechnology*; IntechOpen: London, UK, 2016. [CrossRef]
28. Philippova, O.; Barabanova, A.; Molchanov, V.; Khokhlov, A. Magnetic polymer beads: Recent trends and developments in synthetic design and applications. *Eur. Polym. J.* **2011**, *47*, 542–559. [CrossRef]
29. Khan, I.; Saeed, K.; Khan, I. Nanoparticles: Properties, applications and toxicities. *Arab. J. Chem.* **2019**, *12*, 908–931. [CrossRef]
30. Yurkov, G.Y.; Gubin, S.P.; Ovchenkov, E.A. Magnetic Nanocomposites Based on the Metal-Containing (Fe, Co, Ni) Nanoparticles inside the Polyethylene Matrix. *Magn. Nanoparticles* **2009**, 87–115. [CrossRef]
31. Hanemann, T.; Szabó, D.V. Polymer-Nanoparticle Composites: From Synthesis to Modern Applications. *Materials* **2010**, *3*, 3468–3517. [CrossRef]
32. Alnoch, R.C.; Dos Santos, L.A.; de Almeida, J.M.; Krieger, N.; Mateo, C. Recent trends in biomaterials for immobilization of lipases for application in non-conventional media. *Catalysts* **2020**, *10*, 697. [CrossRef]
33. Bezerra, C.S.; De Farias Lemos, C.M.G.; De Sousa, M.; Gonçalves, L.R.B. Enzyme immobilization onto renewable polymeric matrixes: Past, present, and future trends. *J. Appl. Polym. Sci.* **2015**, *132*, 1–15. [CrossRef]
34. Zdarta, J.; Meyer, A.S.; Jesionowski, T.; Pinelo, M. A general overview of support materials for enzyme immobilization: Characteristics, properties, practical utility. *Catalysts* **2018**, *8*, 92. [CrossRef]
35. Khan, U.S.; Khattak, N.S.; Rahman, A.; Khan, F. Historical development of magnetite nanoparticles synthesis. *J. Chem. Soc. Pakistan* **2011**, *33*, 793–804.
36. Benelmekki, M.; Benelmekki, M. An introduction to nanoparticles and nanotechnology. *Des. Hybrid Nanoparticles* **2014**, *2017*. [CrossRef]

37. Sugimoto, T.; Matijević, E. Formation of uniform spherical magnetite particles by crystallization from ferrous hydroxide gels. *J. Colloid Interface Sci.* **1980**, *74*, 227–243. [CrossRef]
38. Sun, S.; Zeng, H. Size-controlled synthesis of magnetite nanoparticles. *J. Am. Chem. Soc.* **2002**, *124*, 8204–8205. [CrossRef]
39. Hasany, S.F.; Ahmed, I.; Rajan, J.; Rehman, A. Systematic Review of the Preparation Techniques of Iron Oxide Magnetic Nanoparticles. *Nanosci. Nanotechnol.* **2013**, *2*, 148–158. [CrossRef]
40. Jiang, W.; Lai, K.L.; Hu, H.; Zeng, X.B.; Lan, F.; Liu, K.X.; Wu, Y.; Gu, Z.W. The effect of $[Fe^{3+}]/[Fe^{2+}]$ molar ratio and iron salts concentration on the properties of superparamagnetic iron oxide nanoparticles in the water/ethanol/toluene system. *J. Nanoparticle Res.* **2011**, *13*, 5135–5145. [CrossRef]
41. Khurshid, H.; Li, W.; Chandra, S.; Phan, M.H.; Hadjipanayis, G.C.; Mukherjee, P.; Srikanth, H. Mechanism and controlled growth of shape and size variant core/shell FeO/Fe_3O_4 nanoparticles. *Nanoscale* **2013**, *5*, 7942–7952. [CrossRef]
42. Fried, T.; Shemer, G.; Markovich, G. Ordered two-dimensional arrays of ferrite nanoparticles. *Adv. Mater.* **2001**, *13*, 1158–1161. [CrossRef]
43. Kassabova-Zhetcheva, V.D.; Pavlova, L.P.; Samuneva, B.I.; Cherkezova-Zheleva, Z.P.; Mitov, I.G.; Mikhov, M.T. Characterization of superparamagnetic MgxZn1-x Fe_2O_4 powders. *Cent. Eur. J. Chem.* **2007**, *5*, 107–117. [CrossRef]
44. Nemati, Z.; Alonso, J.; Martinez, L.M.; Khurshid, H.; Garaio, E.; Garcia, J.A.; Phan, M.H.; Srikanth, H. Enhanced Magnetic Hyperthermia in Iron Oxide Nano-Octopods: Size and Anisotropy Effects. *J. Phys. Chem. C* **2016**, *120*, 8370–8379. [CrossRef]
45. Alonso, J.; Barandiarán, J.M.; Fernández Barquín, L.; García-Arribas, A. *Magnetic Nanoparticles, Synthesis, Properties, and Applications*; El-Gendy, J.M., Barandiarán, J.M., Hadimani, R.L., Eds.; Elsevier: Amsterdam, The Netherlands, 2018; pp. 1–40. ISBN 9780128139059.
46. Solans, C.; Izquierdo, P.; Nolla, J.; Azemar, N.; Garcia-Celma, M.J. Nano-emulsions. *Curr. Opin. Colloid Interface Sci.* **2005**, *10*, 102–110. [CrossRef]
47. Pileni, M.P. Reverse micelles as microreactors. *J. Phys. Chem.* **1993**, *97*, 6961–6973. [CrossRef]
48. Chin, A.B.; Yaacob, I.I. Synthesis and characterization of magnetic iron oxide nanoparticles via w/o microemulsion and Massart's procedure. *J. Mater. Process. Technol.* **2007**, *191*, 235–237. [CrossRef]
49. Biehl, P.; von der Lühe, M.; Dutz, S.; Schacher, F.H. Synthesis, characterization, and applications of magnetic nanoparticles featuring polyzwitterionic coatings. *Polymers* **2018**, *10*, 91. [CrossRef]
50. Baker, I. *Magnetic Nanoparticle Synthesis*; Elsevier Ltd.: Amsterdam, The Netherlands, 2018; ISBN 9780081007167.
51. Tang, B.; Yuan, L.; Shi, T.; Yu, L.; Zhu, Y. Preparation of nano-sized magnetic particles from spent pickling liquors by ultrasonic-assisted chemical co-precipitation. *J. Hazard. Mater.* **2009**, *163*, 1173–1178. [CrossRef]
52. Wang, Y.; Nkurikiyimfura, I.; Pan, Z. Sonochemical Synthesis of Magnetic Nanoparticles. *Chem. Eng. Commun.* **2015**, *202*, 616–621. [CrossRef]
53. Mathew, D.S.; Juang, R.S. An overview of the structure and magnetism of spinel ferrite nanoparticles and their synthesis in microemulsions. *Chem. Eng. J.* **2007**, *129*, 51–65. [CrossRef]
54. Bang, J.H.; Suslick, K.S. Sonochemical Synthesis of Nanosized Hollow Hematite. *J. Am. Chem. Soc.* **2007**, *129*, 2242–2243. [CrossRef]
55. Vijayakumar, R.; Koltypin, Y.; Felner, I.; Gedanken, A. Sonochemical synthesis and characterization of pure nanometer-sized Fe_3O_4 particles. *Mater. Sci. Eng. A* **2000**, *286*, 101–105. [CrossRef]
56. Weng, Y.C.; Rusakova, I.A.; Baikalov, A.; Chen, J.W.; Wu, N.L. Microstructural evolution of nanocrystalline magnetite synthesized by electrocoagulation. *J. Mater. Res.* **2005**, *20*, 75–80. [CrossRef]
57. Cabrera, L.; Gutierrez, S.; Menendez, N.; Morales, M.P.; Herrasti, P. Magnetite nanoparticles: Electrochemical synthesis and characterization. *Electrochim. Acta* **2008**, *53*, 3436–3441. [CrossRef]
58. Ibrahim, M.; Serrano, K.G.; Noe, L.; Garcia, C.; Verelst, M. Electro-precipitation of magnetite nanoparticles: An electrochemical study. *Electrochim. Acta* **2009**, *55*, 155–158. [CrossRef]
59. Franger, S.; Berthet, P.; Berthon, J. Electrochemical synthesis of Fe_3O_4 nanoparticles in alkaline aqueous solutions containing complexing agents. *J. Solid State Electrochem.* **2004**, *8*, 218–223. [CrossRef]
60. Mosivand, S.; Monzon, L.M.A.; Ackland, K.; Kazeminezhad, I.; Coey, J.M.D. The effect of organics on the structure and magnetization of electro-synthesised magnetite nanoparticles. *J. Nanoparticle Res.* **2013**, *15*. [CrossRef]
61. Fajaroh, F.; Setyawan, H.; Sutrisno; Nazriati; Wonorahardjo, S. To enhance the purity and crystallinity of magnetite nanoparticles prepared by surfactant-free electrochemical method by imposing higher voltage. *AIP Conf. Proc.* **2014**, *1586*, 179–182. [CrossRef]
62. Marques, R.F.C.; Garcia, C.; Lecante, P.; Ribeiro, S.J.L.; Noé, L.; Silva, N.J.O.; Amaral, V.S.; Millán, A.; Verelst, M. Electro-precipitation of Fe_3O_4 nanoparticles in ethanol. *J. Magn. Magn. Mater.* **2008**, *320*, 2311–2315. [CrossRef]
63. Marín, T.; Ortega, D.; Montoya, P.; Arnache, O.; Calderón, J. A new contribution to the study of the electrosynthesis of magnetic nanoparticles: The influence of the supporting electrolyte. *J. Appl. Electrochem.* **2014**, *44*, 1401–1410. [CrossRef]
64. Xu, J.K.; Zhang, F.F.; Sun, J.J.; Sheng, J.; Wang, F.; Sun, M. Bio and nanomaterials based on Fe_3O_4. *Molecules* **2014**, *19*, 21506–21528. [CrossRef]
65. Xia, T.; Wang, J.; Wu, C.; Meng, F.; Shi, Z.; Lian, J.; Feng, J.; Meng, J. Novel complex-coprecipitation route to form high quality triethanolamine-coated Fe_3O_4 nanocrystals: Their high saturation magnetizations and excellent water treatment properties. *CrystEngComm* **2012**, *14*, 5741–5744. [CrossRef]

66. Muthiah, M.; Park, I.K.; Cho, C.S. Surface modification of iron oxide nanoparticles by biocompatible polymers for tissue imaging and targeting. *Biotechnol. Adv.* **2013**, *31*, 1224–1236. [CrossRef]
67. Gautam, A.; Van Veggel, F.C.J.M. Synthesis of nanoparticles, their biocompatibility, and toxicity behavior for biomedical applications. *J. Mater. Chem. B* **2013**, *1*, 5186–5200. [CrossRef] [PubMed]
68. Kydralieva, K.A.; Dzhardimalieva, G.I.; Yurishcheva, A.A.; Jorobekova, S.J. Nanoparticles of Magnetite in Polymer Matrices: Synthesis and Properties. *J. Inorg. Organomet. Polym. Mater.* **2016**, *26*, 1212–1230. [CrossRef]
69. Arias, L.S.; Pessan, J.P.; Vieira, A.P.M.; De Lima, T.M.T.; Delbem, A.C.B.; Monteiro, D.R. Iron oxide nanoparticles for biomedical applications: A perspective on synthesis, drugs, antimicrobial activity, and toxicity. *Antibiotics* **2018**, *7*, 46. [CrossRef] [PubMed]
70. Sahoo, Y.; Pizem, H.; Fried, T.; Golodnitsky, D.; Burstein, L.; Sukenik, C.N.; Markovich, G.; Gan, R. Alkyl Phosphonate/Phosphate Coating on Magnetite Nanoparticles: A Comparison with Fatty Acids. *Langmuir* **2001**, *17*, 7907–7911. [CrossRef]
71. Wang, Y.; Wong, J.F.; Teng, X.; Lin, X.Z.; Yang, H. "Pulling" Nanoparticles into Water: Phase Transfer of Oleic Acid Stabilized Monodisperse Nanoparticles into Aqueous Solutions of α-Cyclodextrin. *Nano Lett.* **2003**, *3*, 1555–1559. [CrossRef]
72. Luchini, A.; Heenan, R.K.; Paduano, L.; Vitiello, G. Functionalized SPIONs: The surfactant nature modulates the self-assembly and cluster formation. *Phys. Chem. Chem. Phys.* **2016**, *18*, 18441–18449. [CrossRef]
73. Wu, W.; Wu, Z.; Yu, T.; Jiang, C.; Kim, W.S. Recent progress on magnetic iron oxide nanoparticles: Synthesis, surface functional strategies and biomedical applications. *Sci. Technol. Adv. Mater.* **2015**, *16*, 23501. [CrossRef]
74. Zhu, N.; Ji, H.; Yu, P.; Niu, J.; Farooq, M.U.; Akram, M.W.; Udego, I.O.; Li, H.; Niu, X. Surface modification of magnetic iron oxide nanoparticles. *Nanomaterials* **2018**, *8*, 810. [CrossRef]
75. Wang, J.; Zhou, H.; Zhuang, J.; Liu, Q. Magnetic γ-Fe_2O_3, Fe_3O_4, and Fe nanoparticles confined within ordered mesoporous carbons as efficient microwave absorbers. *Phys. Chem. Chem. Phys.* **2015**, *17*, 3802–3812. [CrossRef]
76. Liu, X.; Ma, Y.; Zhang, Q.; Zheng, Z.; Wang, L.-S.; Peng, D.L. Facile synthesis of Fe_3O_4/C composites for broadband microwave absorption properties. *Appl. Surf. Sci.* **2018**, *445*, 82–88. [CrossRef]
77. Lu, A.H.; Salabas, E.L.; Schüth, F. Magnetic nanoparticles: Synthesis, protection, functionalization, and application. *Angew. Chemie Int. Ed.* **2007**, *46*, 1222–1244. [CrossRef] [PubMed]
78. Gupta, A.K.; Gupta, M. Synthesis and surface engineering of iron oxide nanoparticles for biomedical applications. *Biomaterials* **2005**, *26*, 3995–4021. [CrossRef] [PubMed]
79. Wagner, J.; Autenrieth, T.; Hempelmann, R. Core shell particles consisting of cobalt ferrite and silica as model ferrofluids [$CoFe_2O_4$-SiO_2 core shell particles]. *J. Magn. Magn. Mater.* **2002**, *252*, 4–6. [CrossRef]
80. Couto, D.; Freitas, M.; Carvalho, F.; Fernandes, E. Iron Oxide Nanoparticles: An Insight into their Biomedical Applications. *Curr. Med. Chem.* **2015**, *22*, 1808–1828. [CrossRef] [PubMed]
81. Costa, C.; Brandão, F.; Bessa, M.J.; Costa, S.; Valdiglesias, V.; Kiliç, G.; Fernández-Bertólez, N.; Quaresma, P.; Pereira, E.; Pásaro, E.; et al. In vitro cytotoxicity of superparamagnetic iron oxide nanoparticles on neuronal and glial cells. Evaluation of nanoparticle interference with viability tests. *J. Appl. Toxicol.* **2016**, *36*, 361–372. [CrossRef]
82. Raghunath, A.; Perumal, E. Metal oxide nanoparticles as antimicrobial agents: A promise for the future. *Int. J. Antimicrob. Agents* **2017**, *49*, 137–152. [CrossRef]
83. Thorek, D.L.J.; Tsourkas, A. Size, charge and concentration dependent uptake of iron oxide particles by non-phagocytic cells. *Biomaterials* **2008**, *29*, 3583–3590. [CrossRef]
84. Zhu, X.M.; Wang, Y.X.J.; Leung, K.C.F.; Lee, S.F.; Zhao, F.; Wang, D.W.; Lai, J.M.Y.; Wan, C.; Cheng, C.H.K.; Ahuja, A.T. Enhanced cellular uptake of aminosilane-coated superparamagnetic iron oxide nanoparticles in mammalian cell lines. *Int. J. Nanomed.* **2012**, *7*, 953–964. [CrossRef]
85. Ling, W.; Wang, M.; Xiong, C.; Xie, D.; Chen, Q.; Chu, X.; Qiu, X.; Li, Y.; Xiao, X. Synthesis, surface modification, and applications of magnetic iron oxide nanoparticles. *J. Mater. Res.* **2019**, *34*, 1828–1844. [CrossRef]
86. Lee, M.H.; Leu, C.C.; Lin, C.C.; Tseng, Y.F.; Lin, H.Y.; Yang, C.N. Gold-decorated magnetic nanoparticles modified with hairpin-shaped DNA for fluorometric discrimination of single-base mismatch DNA. *Microchim. Acta* **2019**, *186*. [CrossRef]
87. Wu, W.; Jiang, C.; Roy, V.A.L. Recent progress in magnetic iron oxide-semiconductor composite nanomaterials as promising photocatalysts. *Nanoscale* **2015**, *7*, 38–58. [CrossRef] [PubMed]
88. Wu, W.; Xiao, X.; Zhang, S.; Ren, F.; Jiang, C. Facile method to synthesize magnetic iron oxides/TiO_2 hybrid nanoparticles and their photodegradation application of methylene blue. *Nanoscale Res. Lett.* **2011**, *6*, 1–15. [CrossRef] [PubMed]
89. Xu, H.; Ouyang, S.; Liu, L.; Reunchan, P.; Umezawa, N.; Ye, J. Recent advances in TiO2-based photocatalysis. *J. Mater. Chem. A* **2014**, *2*, 12642–12661. [CrossRef]
90. Wu, W.; Zhang, S.; Ren, F.; Xiao, X.; Zhou, J.; Jiang, C. Controlled synthesis of magnetic iron oxides@SnO 2 quasi-hollow core-shell heterostructures: Formation mechanism, and enhanced photocatalytic activity. *Nanoscale* **2011**, *3*, 4676–4684. [CrossRef]
91. Li, S.K.; Huang, F.Z.; Wang, Y.; Shen, Y.H.; Qiu, L.G.; Xie, A.J.; Xu, S.J. Magnetic Fe_3O_4@C@Cu_2O composites with bean-like core/shell nanostructures: Synthesis, properties and application in recyclable photocatalytic degradation of dye pollutants. *J. Mater. Chem.* **2011**, *21*, 7459–7466. [CrossRef]
92. Saffari, J.; Mir, N.; Ghanbari, D.; Khandan-Barani, K.; Hassanabadi, A.; Hosseini-Tabatabaei, M.R. Sonochemical synthesis of Fe_3O_4/ZnO magnetic nanocomposites and their application in photo-catalytic degradation of various organic dyes. *J. Mater. Sci. Mater. Electron.* **2015**, *26*, 9591–9599. [CrossRef]

93. Joseph, J.; Nishad, K.K.; Sharma, M.; Gupta, D.K.; Singh, R.R.; Pandey, R.K. Fe_3O_4 and CdS based bifunctional core-shell nanostructure. *Mater. Res. Bull.* **2012**, *47*, 1471–1477. [CrossRef]
94. Liu, L.; Xiao, L.; Zhu, H.Y.; Shi, X.W. Studies on interaction and illumination damage of CS-Fe_3O_4@ZnS:Mn to bovine serum albumin. *J. Nanoparticle Res.* **2013**, *15*. [CrossRef]
95. Zhou, W.; Chen, Y.; Wang, X.; Gu, Z.; Hu, Y. Synthesis of Fe_3O_4@PbS hybrid nanoparticles through the combination of surface-initiated atom transfer radical polymerization and acidolysis by H 2S. *J. Nanosci. Nanotechnol.* **2011**, *11*, 98–105. [CrossRef] [PubMed]
96. Luo, S.; Chai, F.; Zhang, L.; Wang, C.; Li, L.; Liu, X.; Su, Z. Facile and fast synthesis of urchin-shaped Fe_3O_4@Bi_2S_3 core-shell hierarchical structures and their magnetically recyclable photocatalytic activity. *J. Mater. Chem.* **2012**, *22*, 4832–4836. [CrossRef]
97. Rădulescu, M.; Andronescu, E.; Holban, A.M.; Vasile, B.S.; Iordache, F.; Mogoantă, L.; Dan Mogoșanu, G.; Grumezescu, A.M.; Georgescu, M.; Chifiriuc, M.C. Antimicrobial nanostructured bioactive coating based on Fe_3O_4 and patchouli oil for wound dressing. *Metals* **2016**, *6*, 103. [CrossRef]
98. Mu, Q.; Kievit, F.M.; Kant, R.J.; Lin, G.; Jeon, M.; Zhang, M. Anti-HER2/neu peptide-conjugated iron oxide nanoparticles for targeted delivery of paclitaxel to breast cancer cells. *Nanoscale* **2015**, *7*, 18010–18014. [CrossRef] [PubMed]
99. Jahanban-Esfahlan, A.; Dastmalchi, S.; Davaran, S. A simple improved desolvation method for the rapid preparation of albumin nanoparticles. *Int. J. Biol. Macromol.* **2016**, *91*, 703–709. [CrossRef] [PubMed]
100. Nosrati, H.; Sefidi, N.; Sharafi, A.; Danafar, H.; Manjili, H.K. Bovine Serum Albumin (BSA) coated iron oxide magnetic nanoparticles as biocompatible carriers for curcumin-anticancer drug. *Bioorg. Chem.* **2018**, *76*, 501–509. [CrossRef] [PubMed]
101. McBain, S.C.; Yiu, H.H.P.; Dobson, J. Magnetic nanoparticles for gene and drug delivery. *Int. J. Nanomed.* **2008**, *3*, 169–180. [CrossRef] [PubMed]
102. Kharisov, B.I.; Eldin, M.S.M. Enzyme Immobilization: Nanopolymers for Enzyme Immobilization Applications. *CRC Concise Encycl. Nanotechnol.* **2018**, 220–228. [CrossRef]
103. Nelson, J.M.; Griffin, E.G. Adsorption of invertase. *J. Am. Chem. Soc.* **1916**, *38*, 1109–1115. [CrossRef]
104. Mohamad, N.R.; Marzuki, N.H.C.; Buang, N.A.; Huyop, F.; Wahab, R.A. An overview of technologies for immobilization of enzymes and surface analysis techniques for immobilized enzymes. *Biotechnol. Biotechnol. Equip.* **2015**, *29*, 205–220. [CrossRef]
105. Cao, L. Carrier-bound Immobilized Enzymes: Principles, Application and Design. *Carrier-bound Immobil. Enzym. Princ. Appl. Des.* **2006**, 1–563. [CrossRef]
106. Homaei, A.A.; Sariri, R.; Vianello, F.; Stevanato, R. Enzyme immobilization: An update. *J. Chem. Biol.* **2013**, *6*, 185–205. [CrossRef]
107. Ahmad, R.; Sardar, M. Enzyme Immobilization: An Overview on Nanoparticles as Immobilization Matrix. *Biochem. Anal. Biochem.* **2015**, *4*. [CrossRef]
108. Polshettiwar, V.; Luque, R.; Fihri, A.; Zhu, H.; Bouhrara, M.; Basset, J.M. Magnetically recoverable nanocatalysts. *Chem. Rev.* **2011**, *111*, 3036–3075. [CrossRef] [PubMed]
109. Nisha, S.; Karthick, S.A.; Gobi, N. A Review on Methods, Application and Properties of Immobilized Enzyme. *Chem. Sci. Rev. Lett.* **2012**, *1*, 148–155.
110. Rao, S.V.; Anderson, K.W.; Bachas, L.G. Oriented Immobilization of Proteins. *Mikrochim. Acta* **1998**, *128*, 127–143. [CrossRef]
111. Lee, C.H.; Lin, T.S.; Mou, C.Y. Mesoporous materials for encapsulating enzymes. *Nano Today* **2009**, *4*, 165–179. [CrossRef]
112. Yiitolu, M.; Temoçin, Z. Immobilization of Candida rugosa lipase on glutaraldehyde-activated polyester fiber and its application for hydrolysis of some vegetable oils. *J. Mol. Catal. B Enzym.* **2010**, *66*, 130–135. [CrossRef]
113. Lai, B.H.; Yeh, C.C.; Chen, D.H. Surface modification of iron oxide nanoparticles with polyarginine as a highly positively charged magnetic nano-adsorbent for fast and effective recovery of acid proteins. *Process Biochem.* **2012**, *47*, 799–805. [CrossRef]
114. Thangaraj, B.; Solomon, P.R. Immobilization of Lipases–A Review. Part I: Enzyme Immobilization. *ChemBioEng Rev.* **2019**, *6*, 157–166. [CrossRef]
115. Wickramathilaka, M.P.; Tao, B.Y. Characterization of covalent crosslinking strategies for synthesizing DNA-based bioconjugates. *J. Biol. Eng.* **2019**, *13*, 8–17. [CrossRef]
116. Vashist, S.K.; Zhang, B.; Zheng, D.; Al-Rubeaan, K.; Luong, J.H.T.; Sheu, F.S. Sulfo-N-hydroxysuccinimide interferes with bicinchoninic acid protein assay. *Anal. Biochem.* **2011**, *417*, 156–158. [CrossRef]
117. Bart, J.; Tiggelaar, R.; Yang, M.; Schlautmann, S.; Zuilhof, H.; Gardeniers, H. Room-temperature intermediate layer bonding for microfluidic devices. *Lab Chip* **2009**, *9*, 3481–3488. [CrossRef] [PubMed]
118. Eş, I.; Vieira, J.D.G.; Amaral, A.C. Principles, techniques, and applications of biocatalyst immobilization for industrial application. *Appl. Microbiol. Biotechnol.* **2015**, *99*, 2065–2082. [CrossRef] [PubMed]
119. Chakraborty, S.; Rusli, H.; Nath, A.; Sikder, J.; Bhattacharjee, C.; Curcio, S.; Drioli, E. Immobilized biocatalytic process development and potential application in membrane separation: A review. *Crit. Rev. Biotechnol.* **2016**, *36*, 43–48. [CrossRef] [PubMed]
120. Bhatia, S. *Natural Polymer Drug Delivery Systems: Nanoparticles, Plants, and Algae*; Springer International Publishing: Cham, Switzerland, 2016; pp. 1–225. [CrossRef]
121. Mohiti-Asli, M.; Loboa, E.G. *Nanofibrous Smart Bandages for Wound Care*; Elsevier Ltd.: Amsterdam, The Netherlands, 2016; Volume 2, ISBN 9780081006061.
122. Kim, J.E.; Shin, J.Y.; Cho, M.H. Magnetic nanoparticles: An update of application for drug delivery and possible toxic eVects. *Arch. Toxicol.* **2012**, *86*, 685–700. [CrossRef]

123. Tai, M.F.; Lai, C.W.; Abdul Hamid, S.B. Facile Synthesis Polyethylene Glycol Coated Magnetite Nanoparticles for High Colloidal Stability. *J. Nanomater.* **2016**, *2016*. [CrossRef]
124. Mukhopadhyay, A.; Joshi, N.; Chattopadhyay, K.; De, G. A facile synthesis of PEG-coated magnetite (Fe_3O_4) nanoparticles and their prevention of the reduction of cytochrome C. *ACS Appl. Mater. Interfaces* **2012**, *4*, 142–149. [CrossRef]
125. Chang, M.; Chang, Y.J.; Chao, P.Y.; Yu, Q. Exosome purification based on peg-coated Fe_3O_4 Nanoparticles. *PLoS ONE* **2018**, *13*, 1–13. [CrossRef]
126. Hoshino, A.; Costa-Silva, B.; Shen, T.L.; Rodrigues, G.; Hashimoto, A.; Mark, M.T.; Molina, H.; Kohsaka, S.; Di Giannatale, A.; Ceder, S.; et al. Tumour exosome integrins determine organotropic metastasis. *Nature* **2015**, *527*, 329–335. [CrossRef]
127. Zhang, X.; Yuan, X.; Shi, H.; Wu, L.; Qian, H.; Xu, W. Exosomes in cancer: Small particle, big player. *J. Hematol. Oncol.* **2015**, *8*, 1–13. [CrossRef]
128. Rahayu, L.B.H.; Wulandari, I.O.; Santjojo, D.H.; Sabarudin, A. Synthesis and Characterization of Fe_3O_4 Nanoparticles with Polyvinyl Alcohol (PVA) as Capping Agent and Glutaraldehyde as Crosslinkers. *Nature B* **2018**, *4*, 127–134. [CrossRef]
129. Hassan, C.M.; Peppas, N.A. Structure and applications of poly(vinyl alcohol) hydrogels produced by conventional crosslinking or by freezing/thawing methods. *Adv. Polym. Sci.* **2000**, *153*, 37–65. [CrossRef]
130. Nadeem, M.; Ahmad, M.; Akhtar, M.S.; Shaari, A.; Riaz, S.; Naseem, S.; Masood, M.; Saeed, M.A. Magnetic properties of polyvinyl alcohol and doxorubicine loaded iron oxide nanoparticles for anticancer drug delivery applications. *PLoS ONE* **2016**, *11*, 1–12. [CrossRef] [PubMed]
131. Mahmoudi, M.; Shokrgozar, M.A.; Sardari, S.; Moghadam, M.K.; Vali, H.; Laurent, S.; Stroeve, P. Irreversible changes in protein conformation due to interaction with superparamagnetic iron oxide nanoparticles. *Nanoscale* **2011**, *3*, 1127–1138. [CrossRef] [PubMed]
132. Laochai, T.; Mooltongchun, M.; Teepoo, S. Design and Construction of Magnetic Nanoparticles Incorporated with a Chitosan and Poly (vinyl) Alcohol Cryogel and its Application for Immobilization of Horseradish Peroxidase. *Energy Procedia* **2016**, *89*, 248–254. [CrossRef]
133. Ali, M.; Tahir, M.N.; Siwy, Z.; Neumann, R.; Tremel, W.; Ensinger, W. Hydrogen peroxide sensing with horseradish peroxidase-modified polymer single conical nanochannels. *Anal. Chem.* **2011**, *83*, 1673–1680. [CrossRef]
134. Sahin, S.; Ozmen, I. Covalent immobilization of trypsin on polyvinyl alcohol-coated magnetic nanoparticles activated with glutaraldehyde. *J. Pharm. Biomed. Anal.* **2020**, *184*, 113195. [CrossRef]
135. Gerlt, J.A. Stabilization of Reactive Intermediates and Transition States in Enzyme Active Sites by Hydrogen Bonding. *Compr. Nat. Prod. Chem.* **1999**, 5–29. [CrossRef]
136. Lü, J.M.; Wang, X.; Marin-Muller, C.; Wang, H.; Lin, P.H.; Yao, Q.; Chen, C. Current advances in research and clinical applications of PLGA-based nanotechnology. *Expert Rev. Mol. Diagn.* **2011**, *23*, 1–7. [CrossRef]
137. Hirenkumar, M.; Steven, S. Poly Lactic-co-Glycolic Acid (PLGA) as Biodegradable Controlled Drug Delivery Carrier. *Polymers* **2012**, *3*, 1377. [CrossRef]
138. Sun, X.; Xu, C.; Wu, G.; Ye, Q.; Wang, C. Review poly(lactic-co-glycolic acid): Applications and future prospects for periodontal tissue regeneration. *Polymers* **2017**, *9*, 189. [CrossRef]
139. Danhier, F.; Ansorena, E.; Silva, J.M.; Coco, R.; Le Breton, A.; Préat, V. PLGA-based nanoparticles: An overview of biomedical applications. *J. Control. Release* **2012**, *161*, 505–522. [CrossRef] [PubMed]
140. Ganipineni, L.P.; Ucakar, B.; Joudiou, N.; Bianco, J.; Danhier, P.; Zhao, M.; Bastiancich, C.; Gallez, B.; Danhier, F.; Préat, V. Magnetic targeting of paclitaxel-loaded poly(lactic-co-glycolic acid)-based nanoparticles for the treatment of glioblastoma. *Int. J. Nanomed.* **2018**, *13*, 4509–4521. [CrossRef] [PubMed]
141. Spillmann, C.M.; Naciri, J.; Algar, W.R.; Medintz, I.L.; Delehanty, J.B. Multifunctional liquid crystal nanoparticles for intracellular fluorescent imaging and drug delivery. *ACS Nano* **2014**, *8*, 6986–6997. [CrossRef] [PubMed]
142. Key, J.; Leary, J.F. Nanoparticles for multimodal in vivo imaging in nanomedicine. *Int. J. Nanomed.* **2014**, *9*, 711–726. [CrossRef]
143. Yang, J.; Lee, C.H.; Park, J.; Seo, S.; Lim, E.K.; Song, Y.J.; Suh, J.S.; Yoon, H.G.; Huh, Y.M.; Haam, S. Antibody conjugated magnetic PLGA nanoparticles for diagnosis and treatment of breast cancer. *J. Mater. Chem.* **2007**, *17*, 2695–2699. [CrossRef]
144. Thorn, C.F.; Oshiro, C.; Marsh, S.; Hernandez-Boussard, T.; McLeod, H.; Klein, T.E.; Altman, R.B. Doxorubicin pathways: Pharmacodynamics and adverse effects. *Pharm. Genom.* **2011**, *21*, 440–446. [CrossRef] [PubMed]
145. Shubhra, Q.T.H.; Tóth, J.; Gyenis, J.; Feczkó, T. Surface modification of HSA containing magnetic PLGA nanoparticles by poloxamer to decrease plasma protein adsorption. *Colloids Surfaces B Biointerfaces* **2014**, *122*, 529–536. [CrossRef]
146. Lu, W.; Ling, M.; Jia, M.; Huang, P.; Li, C.; Yan, B. Facile synthesis and characterization of polyethylenimine-coated Fe_3O_4 superparamagnetic nanoparticles for cancer cell separation. *Mol. Med. Rep.* **2014**, *9*, 1080–1084. [CrossRef]
147. Wang, X.; Zhou, L.; Ma, Y.; Li, X.; Gu, H. Control of aggregate size of polyethyleneimine-coated magnetic nanoparticles for magnetofection. *Nano Res.* **2009**, *2*, 365–372. [CrossRef]
148. Kwolek, U.; Wójcik, K.; Janiczek, M.; Nowakowska, M.; Kepczynski, M. Synthesis and antibacterial properties of quaternary ammonium derivative of polyethylenimine. *Polimery* **2017**, *62*, 311–315. [CrossRef]
149. Xia, T.T.; Liu, C.Z.; Hu, J.H.; Guo, C. Improved performance of immobilized laccase on amine-functioned magnetic Fe_3O_4 nanoparticles modified with polyethylenimine. *Chem. Eng. J.* **2016**, *295*, 201–206. [CrossRef]
150. Fernández-Fernández, M.; Sanromán, M.Á.; Moldes, D. Recent developments and applications of immobilized laccase. *Biotechnol. Adv.* **2013**, *31*, 1808–1825. [CrossRef] [PubMed]

151. Crestini, C.; Perazzini, R.; Saladino, R. Oxidative functionalisation of lignin by layer-by-layer immobilised laccases and laccase microcapsules. *Appl. Catal. A Gen.* **2010**, *372*, 115–123. [CrossRef]
152. Giardina, P.; Faraco, V.; Pezzella, C.; Piscitelli, A.; Vanhulle, S.; Sannia, G. Laccases: A never-ending story. *Cell. Mol. Life Sci.* **2010**, *67*, 369–385. [CrossRef] [PubMed]
153. Hahn, V.; Meister, M.; Hussy, S.; Cordes, A.; Enderle, G.; Saningong, A.; Schauer, F. Enhanced laccase-mediated transformation of diclofenac and flufenamic acid in the presence of bisphenol A and testing of an enzymatic membrane reactor. *AMB Express* **2018**, *8*. [CrossRef] [PubMed]
154. Zuvin, M.; Kuruoglu, E.; Kaya, V.O.; Unal, O.; Kutlu, O.; Yagci Acar, H.; Gozuacik, D.; Kosar, A. Magnetofection of green fluorescent protein encoding DNA-bearing polyethyleneimine-coated superparamagnetic iron oxide nanoparticles to human breast cancer cells. *ACS Omega* **2019**, *4*, 12366–12374. [CrossRef]
155. Wiogo, H.T.R.; Lim, M.; Bulmus, V.; Gutiérrez, L.; Woodward, R.C.; Amal, R. Insight into serum protein interactions with functionalized magnetic nanoparticles in biological media. *Langmuir* **2012**, *28*, 4346–4356. [CrossRef]
156. Kannan, K.; Mukherjee, J.; Gupta, M.N. Use of polyethyleneimine coated Fe_3O_4 nanoparticles as an ion-exchanger for protein separation. *Sci. Adv. Mater.* **2013**, *5*, 1477–1484. [CrossRef]
157. Gräfe, C.; Weidner, A.; Lühe, M.V.D.; Bergemann, C.; Schacher, F.H.; Clement, J.H.; Dutz, S. Intentional formation of a protein corona on nanoparticles: Serum concentration affects protein corona mass, surface charge, and nanoparticle-cell interaction. *Int. J. Biochem. Cell Biol.* **2016**, *75*, 196–202. [CrossRef]
158. Calatayud, M.P.; Sanz, B.; Raffa, V.; Riggio, C.; Ibarra, M.R.; Goya, G.F. The effect of surface charge of functionalized Fe_3O_4 nanoparticles on protein adsorption and cell uptake. *Biomaterials* **2014**, *35*, 6389–6399. [CrossRef]
159. Roohi, F.; Lohrke, J.; Ide, A.; Schütz, G.; Dassler, K. Studying the effect of particle size and coating type on the blood kinetics of superparamagnetic iron oxide nanoparticles. *Int. J. Nanomed.* **2012**, *7*, 4447–4458. [CrossRef]
160. Xu, Y.; Zhuang, L.; Lin, H.; Shen, H.; Li, J.W. Preparation and characterization of polyacrylic acid coated magnetite nanoparticles functionalized with amino acids. *Thin Solid Films* **2013**, *544*, 368–373. [CrossRef]
161. Sanchez, L.M.; Martin, D.A.; Alvarez, V.A.; Gonzalez, J.S. Polyacrylic acid-coated iron oxide magnetic nanoparticles: The polymer molecular weight influence. *Colloids Surfaces A Physicochem. Eng. Asp.* **2018**, *543*, 28–37. [CrossRef]
162. Shagholani, H.; Ghoreishi, S.M.; Mousazadeh, M. Improvement of interaction between PVA and chitosan via magnetite nanoparticles for drug delivery application. *Int. J. Biol. Macromol.* **2015**, *78*, 130–136. [CrossRef]
163. Hamidreza, S. Synthesis of Nanocomposition of Poly Acrylic Acid/Chitosan Coated-Magnetite Nanoparticles to Investigation of Interaction with BSA and IGG Proteins. *Int. J. Nanomater. Nanotechnol. Nanomed.* **2017**, *3*, 27–33. [CrossRef]
164. Huang, S.H.; Liao, M.H.; Chen, D.H. Fast and efficient recovery of lipase by polyacrylic acid-coated magnetic nano-adsorbent with high activity retention. *Sep. Purif. Technol.* **2006**, *51*, 113–117. [CrossRef]
165. Liao, M.H.; Chen, D.H. Preparation and characterization of a novel magnetic nano-adsorbent. *J. Mater. Chem.* **2002**, *12*, 3654–3659. [CrossRef]
166. Ma, Y.H.; Wu, S.Y.; Wu, T.; Chang, Y.J.; Hua, M.Y.; Chen, J.P. Magnetically targeted thrombolysis with recombinant tissue plasminogen activator bound to polyacrylic acid-coated nanoparticles. *Biomaterials* **2009**, *30*, 3343–3351. [CrossRef]
167. Zhao, T.; Chen, K.; Gu, H. Investigations on the interactions of proteins with polyampholyte-coated magnetite nanoparticles. *J. Phys. Chem. B* **2013**, *117*, 14129–14135. [CrossRef]
168. Xiao, W.; Gu, H.; Li, D.; Chen, D.; Deng, X.; Jiao, Z.; Lin, J. Microwave-assisted synthesis of magnetite nanoparticles for MR blood pool contrast agents. *J. Magn. Magn. Mater.* **2012**, *324*, 488–494. [CrossRef]
169. Woźniak, E.; Špírková, M.; Šlouf, M.; Garamus, V.M.; Šafaříková, M.; Šafařík, I.; Štěpánek, M. Stabilization of aqueous dispersions of poly(methacrylic acid)-coated iron oxide nanoparticles by double hydrophilic block polyelectrolyte poly(ethylene oxide)-block-poly(N-methyl-2-vinylpyridinium iodide). *Colloids Surfaces A Physicochem. Eng. Asp.* **2017**, *514*, 32–37. [CrossRef]
170. Ahmad, H.; Ahmad, A.; Islam, S.S. Magnetic Fe_3O_4@poly(methacrylic acid) particles for selective preconcentration of trace arsenic species. *Microchim. Acta* **2017**, *184*, 2007–2014. [CrossRef]
171. Li, K.; Chen, K.; Wang, Q.; Zhang, Y.; Gan, W. Synthesis of poly(acrylic acid) coated magnetic nanospheres via a multiple polymerization route. *R. Soc. Open Sci.* **2019**, *6*. [CrossRef]
172. Mekseriwattana, W.; Srisuk, S.; Kriangsaksri, R.; Niamsiri, N.; Prapainop, K. The Impact of Serum Proteins and Surface Chemistry on Magnetic Nanoparticle Colloidal Stability and Cellular Uptake in Breast Cancer Cells. *AAPS PharmSciTech* **2019**, *20*, 1–9. [CrossRef] [PubMed]
173. Yu, S.; Chow, C.M. Carboxyl group (-CO2H) functionalized ferrimagnetic iron oxide nanoparticles for potential bio-applications. *J. Mater. Chem.* **2004**, *14*, 2781–2786. [CrossRef]
174. Lanzalaco, S.; Armelin, E. Poly(*N*-isopropylacrylamide) and Copolymers: A Review on Recent Progresses in Biomedical Applications. *Gels* **2017**, *3*, 36. [CrossRef]
175. Alipour, A.; Shekardasht, M.B.; Gharbani, P. The synthesis, characterization and applications of poly[:*N*-isopropylacrylamide-co-3-allyloxy-1,2-propanediol] grafted onto modified magnetic nanoparticles. *RSC Adv.* **2020**, *10*, 3511–3519. [CrossRef]
176. Nguyen, N.H.A.; Darwish, M.S.A.; Stibor, I.; Kejzlar, P.; Ševců, A. Magnetic Poly(*N*-isopropylacrylamide) Nanocomposites: Effect of Preparation Method on Antibacterial Properties. *Nanoscale Res. Lett.* **2017**, *12*. [CrossRef]
177. Shamim, N.; Hong, L.; Hidajat, K.; Uddin, M.S. Thermosensitive-polymer-coated magnetic nanoparticles: Adsorption and desorption of Bovine Serum Albumin. *J. Colloid Interface Sci.* **2006**, *304*, 1–8. [CrossRef]

178. Dionigi, C.; Lungaro, L.; Goranov, V.; Riminucci, A.; Piñeiro-Redondo, Y.; Bañobre-López, M.; Rivas, J.; Dediu, V. Smart magnetic poly(*N*-isopropylacrylamide) to control the release of bio-active molecules. *J. Mater. Sci. Mater. Med.* **2014**, *25*, 2365–2371. [CrossRef]
179. Cheung, R.C.F.; Ng, T.B.; Wong, J.H.; Chan, W.Y. Chitosan: An Update on Potential Biomedical and Pharmaceutical Applications. *Mar Drugs.* **2015**, *13*, 5156–5186. [CrossRef] [PubMed]
180. Chirkov, S.N. The antiviral activity of chitosan (review). *Appl. Biochem. Microbiol.* **2002**, *38*, 1–8. [CrossRef]
181. Croisier, F.; Jérôme, C. Chitosan-based biomaterials for tissue engineering. *Eur. Polym. J.* **2013**, *49*, 780–792. [CrossRef]
182. Ziegler-Borowska, M.; Siódmiak, T.; Chełminiak, D.; Cyganiuk, A.; Marszałł, M.P. Magnetic nanoparticles with surfaces modified with chitosan-poly[*N*-benzyl-2-(methacryloxy)-N,N-dimethylethanaminium bromide] for lipase immobilization. *Appl. Surf. Sci.* **2014**, *288*, 641–648. [CrossRef]
183. Siódmiak, T.; Ziegler-Borowska, M.; Marszałł, M.P. Lipase-immobilized magnetic chitosan nanoparticles for kinetic resolution of (R,S)-ibuprofen. *J. Mol. Catal. B Enzym.* **2013**, *94*, 7–14. [CrossRef]
184. Ziegler-Borowska, M.; Mylkie, K.; Kozlowska, M.; Nowak, P.; Chelminiak-Dudkiewicz, D.; Kozakiewicz, A.; Ilnicka, A.; Kaczmarek-Kedziera, A. Effect of geometrical structure, drying, and synthetic method on aminated chitosan-coated magnetic nanoparticles utility for HSA effective immobilization. *Molecules* **2019**, *24*, 1925. [CrossRef]
185. Ziegler-Borowska, M.; Mylkie, K.; Nowak, P.; Rybczynski, P.; Sikora, A.; Chelminiak-Dudkiewicz, D.; Kaczmarek-Kedziera, A. Testing for ketoprofen binding to HSA coated magnetic nanoparticles under normal conditions and after oxidative stress. *Molecules* **2020**, *25*, 1945. [CrossRef]
186. Qiao, T.; Wu, Y.; Jin, J.; Gao, W.; Xie, Q.; Wang, S.; Zhang, Y.; Deng, H. Conjugation of catecholamines on magnetic nanoparticles coated with sulfonated chitosan. *Colloids Surfaces A Physicochem. Eng. Asp.* **2011**, *380*, 169–174. [CrossRef]
187. Park, J.H.; Saravanakumar, G.; Kim, K.; Kwon, I.C. Targeted delivery of low molecular drugs using chitosan and its derivatives. *Adv. Drug Deliv. Rev.* **2010**, *62*, 28–41. [CrossRef]
188. Liang, Y.Y.; Zhang, L.M. Bioconjugation of papain on superparamagnetic nanoparticles decorated with carboxymethylated chitosan. *Biomacromolecules* **2007**, *8*, 1480–1486. [CrossRef]
189. Li, G.Y.; Jiang, Y.R.; Huang, K.L.; Ding, P.; Yao, L.L. Kinetics of adsorption of Saccharomyces cerevisiae mandelated dehydrogenase on magnetic Fe_3O_4-chitosan nanoparticles. *Colloids Surfaces A Physicochem. Eng. Asp.* **2008**, *320*, 11–18. [CrossRef]
190. Wang, S.N.; Zhang, C.R.; Qi, B.K.; Sui, X.N.; Jiang, L.Z.; Li, Y.; Wang, Z.J.; Feng, H.X.; Wang, R.; Zhang, Q.Z. Immobilized alcalase alkaline protease on the magnetic chitosan nanoparticles used for soy protein isolate hydrolysis. *Eur. Food Res. Technol.* **2014**, *239*, 1051–1059. [CrossRef]
191. Sojitra, U.V.; Nadar, S.S.; Rathod, V.K. Immobilization of pectinase onto chitosan magnetic nanoparticles by macromolecular cross-linker. *Carbohydr. Polym.* **2017**, *157*, 677–685. [CrossRef] [PubMed]
192. Ziegler-Borowska, M.; Chelminiak-Dudkiewicz, D.; Siódmiak, T.; Sikora, A.; Wegrzynowska-Drzymalska, K.; Skopinska-Wisniewska, J.; Kaczmarek, H.; Marszałł, M.P. Chitosan–collagen coated magnetic nanoparticles for lipase immobilization—new type of "enzyme friendly" polymer shell crosslinking with squaric acid. *Catalysts* **2017**, *7*, 26. [CrossRef]
193. Sikora, A.; Chełminiak-Dudkiewicz, D.; Siódmiak, T.; Tarczykowska, A.; Sroka, W.D.; Ziegler-Borowska, M.; Marszałł, M.P. Enantioselective acetylation of (R,S)-atenolol: The use of Candida rugosa lipases immobilized onto magnetic chitosan nanoparticles in enzyme-catalyzed biotransformation. *J. Mol. Catal. B Enzym.* **2016**, *134*, 43–50. [CrossRef]
194. Hosseini, S.M.; Kim, S.M.; Sayed, M.; Younesi, H.; Bahramifar, N.; Park, J.H.; Pyo, S.H. Lipase-immobilized chitosan-crosslinked magnetic nanoparticle as a biocatalyst for ring opening esterification of itaconic anhydride. *Biochem. Eng. J.* **2019**, *143*, 141–150. [CrossRef]
195. Kuo, C.H.; Liu, Y.C.; Chang, C.M.J.; Chen, J.H.; Chang, C.; Shieh, C.J. Optimum conditions for lipase immobilization on chitosan-coated Fe_3O_4 nanoparticles. *Carbohydr. Polym.* **2012**, *87*, 2538–2545. [CrossRef]
196. Wang, X.Y.; Jiang, X.P.; Li, Y.; Zeng, S.; Zhang, Y.W. Preparation Fe_3O_4@chitosan magnetic particles for covalent immobilization of lipase from Thermomyces lanuginosus. *Int. J. Biol. Macromol.* **2015**, *75*, 44–50. [CrossRef]
197. Monteiro, R.R.C.; Lima, P.J.M.; Pinheiro, B.B.; Freire, T.M.; Dutra, L.M.U.; Fechine, P.B.A.; Gonçalves, L.R.B.; de Souza, M.C.M.; Dos Santos, J.C.S.; Fernandez-Lafuente, R. Immobilization of lipase a from Candida antarctica onto Chitosan-coated magnetic nanoparticles. *Int. J. Mol. Sci.* **2019**, *20*, 4018. [CrossRef]
198. Ziegler-Borowska, M.; Chełminiak, D.; Siódmiak, T.; Sikora, A.; Marszałł, M.P.; Kaczmarek, H. Synthesis of new chitosan coated magnetic nanoparticles with surface modified with long-distanced amino groups as a support for bioligands binding. *Mater. Lett.* **2014**, *132*, 63–65. [CrossRef]
199. Ziegler-Borowska, M.; Chełminiak, D.; Kaczmarek, H.; Kaczmarek-Kędziera, A. Effect of side substituents on thermal stability of the modified chitosan and its nanocomposites with magnetite. *J. Therm. Anal. Calorim.* **2016**, *124*, 1267–1280. [CrossRef]
200. Yi, S.S.; Noh, J.M.; Lee, Y.S. Amino acid modified chitosan beads: Improved polymer supports for immobilization of lipase from Candida rugosa. *J. Mol. Catal. B Enzym.* **2009**, *57*, 123–129. [CrossRef]
201. Sikora, A.; Chełminiak-Dudkiewicz, D.; Ziegler-Borowska, M.; Marszałł, M.P. Enantioseparation of (RS)-atenolol with the use of lipases immobilized onto new-synthesized magnetic nanoparticles. *Tetrahedron Asymmetry* **2017**, *28*, 374–380. [CrossRef]
202. Marszałł, M.P.; Sroka, W.D.; Sikora, A.; Chełminiak, D.; Ziegler-Borowska, M.; Siódmiak, T.; Moaddel, R. Ligand fishing using new chitosan based functionalized Androgen Receptor magnetic particles. *J. Pharm. Biomed. Anal.* **2016**, *127*, 129–135. [CrossRef] [PubMed]

203. Zamora-Mora, V.; Soares, P.; Echeverria, C.; Hernández, R.; Mijangos, C. Composite Chitosan/Agarose Ferrogels for Potential Applications in Magnetic Hyperthermia. *Gels* **2015**, *1*, 69–80. [CrossRef] [PubMed]
204. Gu, Y.; Cheong, K.L.; Du, H. Modification and comparison of three Gracilaria spp. agarose with methylation for promotion of its gelling properties. *Chem. Cent. J.* **2017**, *11*, 1–10. [CrossRef]
205. Adivi, F.G.; Hashemi, P. Ultrafine agarose-coated superparamagnetic iron oxide nanoparticles (AC-SPIONs): A promising sorbent for drug delivery applications. *J. Iran. Chem. Soc.* **2018**, *15*, 1145–1152. [CrossRef]
206. Adivi, F.G.; Hashemi, P.; Tehrani, A.D. Agarose-coated Fe_3O_4 @SiO_2 magnetic nanoparticles modified with sodium dodecyl sulfate, a new promising sorbent for fast adsorption/desorption of cationic drugs. *Polym. Bull.* **2019**, *76*, 1239–1256. [CrossRef]
207. Liao, J.; Huang, H. Review on Magnetic Natural Polymer Constructed Hydrogels as Vehicles for Drug Delivery. *Biomacromolecules* **2020**. [CrossRef]
208. Chen, T.; Yang, W.; Guo, Y.; Yuan, R.; Xu, L.; Yan, Y. Enhancing catalytic performance of β-glucosidase via immobilization on metal ions chelated magnetic nanoparticles. *Enzyme Microb. Technol.* **2014**, *63*, 50–57. [CrossRef]
209. Klimaviciute, R.; Bendoraitiene, J.; Lekniute, E.; Zemaitaitis, A. Non-stoichiometric complexes of cationic starch and 4-sulfophthalic acid and their flocculation efficiency. *Colloids Surfaces A Physicochem. Eng. Asp.* **2014**, *457*, 180–188. [CrossRef]
210. Uthaman, S.; Lee, S.J.; Cherukula, K.; Cho, C.S.; Park, I.K. Polysaccharide-coated magnetic nanoparticles for imaging and gene therapy. *Biomed Res. Int.* **2015**, *2015*. [CrossRef]
211. Adak, S.; Banerjee, R. A green approach for starch modification: Esterification by lipase and novel imidazolium surfactant. *Carbohydr. Polym.* **2016**, *150*, 359–368. [CrossRef] [PubMed]
212. Ziegler-Borowska, M. Magnetic nanoparticles coated with aminated starch for HSA immobilization- simple and fast polymer surface functionalization. *Int. J. Biol. Macromol.* **2019**, *136*, 106–114. [CrossRef] [PubMed]
213. Saikia, C.; Hussain, A.; Ramteke, A.; Sharma, H.K.; Deb, P.; Maji, T.K. Carboxymethyl starch-coated iron oxide magnetic nanoparticles: A potential drug delivery system for isoniazid. *Iran. Polym. J.* **2015**, *24*, 815–828. [CrossRef]
214. Dung, T.T.; Danh, T.M.; Hoa, L.T.M.; Chien, D.M.; Duc, N.H. Structural and magnetic properties of starch-coated magnetite nanoparticles. *J. Exp. Nanosci.* **2009**, *4*, 259–267. [CrossRef]
215. Zheng, M.; Lu, J.; Zhao, D. Effects of starch-coating of magnetite nanoparticles on cellular uptake, toxicity and gene expression profiles in adult zebrafish. *Sci. Total Environ.* **2018**, *622–623*, 930–941. [CrossRef] [PubMed]
216. Wang, B.; Cheng, F.; Lu, Y.; Ge, W.; Zhang, M.; Yue, B. Immobilization of pectinase from Penicillium oxalicum F67 onto magnetic cornstarch microspheres: Characterization and application in juice production. *J. Mol. Catal. B Enzym.* **2013**, *97*, 137–143. [CrossRef]
217. Gagnon, P.; Toh, P.; Lee, J. High productivity purification of immunoglobulin G monoclonal antibodies on starch-coated magnetic nanoparticles by steric exclusion of polyethylene glycol. *J. Chromatogr. A* **2014**, *1324*, 171–180. [CrossRef] [PubMed]
218. Lu, W.; Shen, Y.; Xie, A.; Zhang, W. Preparation and protein immobilization of magnetic dialdehyde starch nanoparticles. *J. Phys. Chem. B* **2013**, *117*, 3720–3725. [CrossRef] [PubMed]
219. Kuroiwa, T.; Noguchi, Y.; Nakajima, M.; Sato, S.; Mukataka, S.; Ichikawa, S. Production of chitosan oligosaccharides using chitosanase immobilized on amylose-coated magnetic nanoparticles. *Process Biochem.* **2008**, *43*, 62–69. [CrossRef]
220. Eyley, S.; Thielemans, W. Surface modification of cellulose nanocrystals. *Nanoscale* **2014**, *6*, 7764–7779. [CrossRef] [PubMed]
221. Habibi, Y.; Lucia, L.A.; Rojas, O.J. Cellulose nanocrystals: Chemistry, self-assembly, and applications. *Chem. Rev.* **2010**, *110*, 3479–3500. [CrossRef]
222. Van Rie, J.; Thielemans, W. Cellulose-gold nanoparticle hybrid materials. *Nanoscale* **2017**, *9*, 8525–8554. [CrossRef] [PubMed]
223. Aguilera, G.; Berry, C.C.; West, R.M.; Gonzalez-Monterrubio, E.; Angulo-Molina, A.; Arias-Carrión, Ó.; Méndez-Rojas, M.Á. Carboxymethyl cellulose coated magnetic nanoparticles transport across a human lung microvascular endothelial cell model of the blood-brain barrier. *Nanoscale Adv.* **2019**, *1*, 671–685. [CrossRef]
224. Namdeo, M.; Bajpai, S.K. Immobilization of α-amylase onto cellulose-coated magnetite (CCM) nanoparticles and preliminary starch degradation study. *J. Mol. Catal. B Enzym.* **2009**, *59*, 134–139. [CrossRef]
225. Ivanova, V.; Petrova, P.; Hristov, J. Application in the Ethanol Fermentation of Immobilized Yeast Cells in Matrix of Alginate/Magnetic Nanoparticles, on Chitosan-Magnetite Microparticles and Cellulose-coated Magnetic Nanoparticles. *arXiv* **2011**, arXiv:1105.0619.
226. Anirudhan, T.S.; Rejeena, S.R.; Binusree, J. Adsorptive separation of myoglobin from aqueous solutions using iron oxide magnetic nanoparticles modified with functionalized nanocrystalline cellulose. *J. Chem. Eng. Data* **2013**, *58*, 1329–1339. [CrossRef]
227. Zhang, J.; Feng, X.; Wang, J.; Fang, G.; Liu, J.; Wang, S. Nano-crystalline cellulose-coated magnetic nanoparticles for affinity adsorption of glycoproteins. *Analyst* **2020**, *145*, 3407–3413. [CrossRef]
228. Mohammadi, F.; Moeeni, M.; Li, C.; Boukherroub, R.; Szunerits, S. Interaction of cellulose and nitrodopamine coated superparamagnetic iron oxide nanoparticles with alpha-lactalbumin. *RSC Adv.* **2020**, *10*, 9704–9716. [CrossRef]
229. Ohannesian, N.; De Leo, C.T.; Martirosyan, K.S. Dextran coated superparamagnetic iron oxide nanoparticles produced by microfluidic process. *Mater. Today Proc.* **2019**, *13*, 397–403. [CrossRef]
230. Unterweger, H.; Dézsi, L.; Matuszak, J.; Janko, C.; Poettler, M.; Jordan, J.; Bäuerle, T.; Szebeni, J.; Fey, T.; Boccaccini, A.R.; et al. Dextran-coated superparamagnetic iron oxide nanoparticles for magnetic resonance imaging: Evaluation of size-dependent imaging properties, storage stability and safety. *Int. J. Nanomed.* **2018**, *13*, 1899–1915. [CrossRef] [PubMed]

231. Naha, P.C.; Liu, Y.; Hwang, G.; Huang, Y.; Gubara, S.; Jonnakuti, V.; Simon-Soro, A.; Kim, D.; Gao, L.; Koo, H.; et al. Dextran-Coated Iron Oxide Nanoparticles as Biomimetic Catalysts for Localized and pH-Activated Biofilm Disruption. *ACS Nano* **2019**, *13*, 4960–4971. [CrossRef] [PubMed]
232. Hradil, J.; Pisarev, A.; Babič, M.; Horák, D. Dextran-modified iron oxide nanoparticles. *China Particuology* **2007**, *5*, 162–168. [CrossRef]
233. Bai, H.; Liu, Z.; Sun, D.D. Highly water soluble and recovered dextran coated fe3o 4 magnetic nanoparticles for brackish water desalination. *Sep. Purif. Technol.* **2011**, *81*, 392–399. [CrossRef]
234. Vasić, K.; Knez, Ž.; Konstantinova, E.A.; Kokorin, A.I.; Gyergyek, S.; Leitgeb, M. Structural and magnetic characteristics of carboxymethyl dextran coated magnetic nanoparticles: From characterization to immobilization application. *React. Funct. Polym.* **2020**, *148*, 104481. [CrossRef]
235. Yu, M.K.; Park, J.; Jon, S. Targeting strategies for multifunctional nanoparticles in cancer imaging and therapy. *Theranostics* **2012**, *2*, 3–44. [CrossRef]
236. Weissleder, R.; Moore, A.; Mahmood, U.; Bhorade, R.; Benveniste, H.; Chiocca, E.A.; Basilion, J.P. In vivo magnetic resonance imaging of transgene expression. *Nat. Med.* **2000**, *6*, 351–354. [CrossRef]
237. Hogemann, D.; Josephson, L.; Weissleder, R.; Basilion, J.P. Improvement of MRI probes to allow efficient detection of gene expression. *Bioconjug. Chem.* **2000**, *11*, 941–946. [CrossRef]
238. Horng, H.E.; Yang, S.Y.; Hong, C.Y.; Liu, C.M.; Tsai, P.S.; Yang, H.C.; Wu, C.C. Biofunctionalized magnetic nanoparticles for high-sensitivity immunomagnetic detection of human C-reactive protein. *Appl. Phys. Lett.* **2006**, *88*, 1–4. [CrossRef]
239. Templin, M.F.; Stoll, D.; Schrenk, M.; Traub, P.C.; Vöhringer, C.F.; Joos, T.O. Protein microarray technology. *Trends Biotechnol.* **2002**, *20*, 160–166. [CrossRef]
240. Li, J.; Zhou, Y.; Li, M.; Xia, N.; Huang, Q.; Do, H.; Liu, Y.N.; Zhou, F. Carboxymethylated dextran-coated magnetic iron oxide nanoparticles for regenerable bioseparation. *J. Nanosci. Nanotechnol.* **2011**, *11*, 10187–10192. [CrossRef] [PubMed]
241. Ziv, O.; Avtalion, R.R.; Margel, S. Immunogenicity of bioactive magnetic nanoparticles: Natural and acquired antibodies. *J. Biomed. Mater. Res. Part A* **2008**, *85*, 1011–1021. [CrossRef] [PubMed]
242. Narain, R.; Gonzales, M.; Hoffman, A.S.; Stayton, P.S.; Krishnan, K.M. Synthesis of monodisperse biotinylated p(NIPAAm)-coated iron oxide magnetic nanoparticles and their bioconjugation to streptavidin. *Langmuir* **2007**, *23*, 6299–6304. [CrossRef] [PubMed]

materials

MDPI

Article

Biodegradable and Drug-Eluting Inorganic Composites Based on Mesoporous Zinc Oxide for Urinary Stent Applications

Marco Laurenti [1], Marta Grochowicz [2], Elena Dragoni [1], Marco Carofiglio [1], Tania Limongi [1] and Valentina Cauda [1,*]

1 Department of Applied Science and Technology, Politecnico di Torino, Corso Duca degli Abruzzi 24, 10129 Torino, Italy; marco.laurenti@polito.it (M.L.); elena.dragoni@studenti.polito.it or elena.dragoni@gmail.com (E.D.); marco.carofiglio@polito.it (M.C.); tania.limongi@polito.it (T.L.)

2 Department of Polymer Chemistry, Institute of Chemical Sciences, Faculty of Chemistry, Maria Curie Sklodowska University, Gliniana 33, 20-614 Lublin, Poland; mgrochowicz@poczta.umcs.lublin.pl or mgrochowicz@umcs.pl

* Correspondence: valentina.cauda@polito.it; Tel.: +39-011-090-7389

Received: 28 July 2020; Accepted: 26 August 2020; Published: 29 August 2020

check for updates

Abstract: Conventional technologies for ureteral stent fabrication suffer from major inconveniences such as the development of encrustations and bacteria biofilm formation. These drawbacks typically lead to the failure of the device, significant patient discomfort and an additional surgery to remove and replace the stent in the worst cases. This work focuses on the preparation of a new nanocomposite material able to show drug elution properties, biodegradation and eventually potential antibacterial activity. Poly(2-hydroxyethyl methacrylate) or the crosslinked poly(2-hydroxyethyl methacrylate)-co-poly(acrylic acid) hydrogels were prepared by the radical polymerization method and combined with a biodegradable and antibacterial filling agent, i.e., flower-like Zinc Oxide (ZnO) micropowders obtained via the hydrothermal route. The physico-chemical analyses revealed the correct incorporation of ZnO within the hydrogel matrix and its highly mesoporous structure and surface area, ideal for drug incorporation. Two different anti-inflammatory drugs (Ibuprofen and Diclofenac) were loaded within each composite and the release profile was monitored up to two weeks in artificial urine (AU) and even at different pH values in AU to simulate pathological conditions. The addition of mesoporous ZnO micropowders to the hydrogel did not negatively affect the drug loading properties of the hydrogel and it was successfully allowed to mitigate undesirable burst-release effects. Furthermore, the sustained release of the drugs over time was observed at neutral pH, with kinetic constants (k) as low as 0.05 h^{-1}. By exploiting the pH-tunable swelling properties of the hydrogel, an even more sustained release was achieved in acidic and alkaline conditions especially at short release times, with a further reduction of burst effects (k ≈ 0.01–0.02 h^{-1}). The nanocomposite system herein proposed represents a new material formulation for preparing innovative drug eluting stents with intrinsic antibacterial properties.

Keywords: mesoporous zinc oxide; polyHEMA; hydrogel; drug-eluting stent; pH-triggered drug release

1. Introduction

Nowadays, the use of ureteral stents is a conventional and routine practice. However, it shows a huge number of drawbacks due to the absence of material optimization and lack of advanced performances, especially in comparison with the advancement obtained in other fields such as cardiovascular stents.

The main limitations of current ureteral stents are (i) the encrustation of the system induced by the precipitation of inorganic salts from urine and (ii) the formation of bacteria biofilm, with consequent antibiotic resistance, persistence of the infections, morbidity, urinary retention, ureteral damage and in the worst cases pyelonephritis and sepsis [1,2]. All these drawbacks lead to significant patient discomfort (pain, urgency, frequency) and a second surgery to remove or replace the stent, even often supported by an alongside antibiotic, analgesic or other drug treatments [3]. As a result of frequent re-hospitalization, an increase in public health costs are also evidenced [4].

In the last decades, several approaches have been investigated to modify the stent itself including a new device design, optimization of the polymeric formulation, the deposition of surface coatings to prevent bacteria and inorganic encrustation and other advanced applications such as drug-elution or complete stent biodegradation to avoid further re-hospitalization [5]. In particular, the use of functional coatings has been widely considered and various materials turned out to be a valid solution for preventing both the adhesion of bacteria and the formation of inorganic encrustation to the stent surface. These included the use of antimicrobial silver [6], hydrogels (like Hydromer®) [7], heparin [8] and diamond-like amorphous carbon coatings [9] to name only a few. On the other side, biodegradable and drug eluting stents made by polylactic acid (PLA) polymers have been tested in animal models with appreciable results [10]. However, the considered stent only partially degraded and small polymer fragments remained in the ureter [11], inducing the obstruction to the urine flow.

Recently, the use of nanostructured multifunctional materials for biomedical applications attracted great attention as these can represent an alternative approach with respect to most of the conventional ones investigated right now [12–14]. Among nanomaterials, Zinc Oxide (ZnO) is one of the most studied [15,16]. ZnO can be used for many applications thanks to its very interesting physical and chemical properties. Moreover, it can be easily prepared in several high-surface-area morphologies by low-cost hydrothermal methods [17,18]. Some of them include nanotubes [19] and nanowires [20,21], nanobelts [22], nanorings [23], flower-like particles [24,25], nanosprings [26], multipods [27,28] and nanoparticles [29]. ZnO is a "generally recognized as safe" (GRAS), white, odorless inorganic solid compound. ZnO also exhibits intrinsic antibacterial properties [30], thereby paving the way to its possible use as intrinsically-antimicrobial material. The antibacterial activity of ZnO is mainly ascribed to the following mechanisms [31]: reactive oxygen species (ROS) generation, release of Zn^{2+} cations, and electrostatic interactions between ZnO nanomaterials and the bacterial surface. Despite the potential role in determining ZnO antibacterial efficacy, the release of Zn^{2+} ions can also induce undesirable cytotoxicity against healthy cells and their release must be properly optimized and monitored to guarantee biocompatibility at the same time [29,32]. Especially for biomedical implants or drug delivery devices, the control over ZnO particle size, morphology, interparticle porosity and surface chemistry is crucial to achieve antibacterial properties and biocompatibility towards healthy cells at the same time [30]. Another important aspect of ZnO is its biodegradability, as it is easily soluble in water solution at pH below 5.5 or in water solution rich in salts, such as phosphates [33]. It is, therefore, clear that the use of ZnO-based materials in the field of ureteral stents is not only novel but can also give many advantages.

To overcome some of the abovementioned limitations, ZnO has also been incorporated into various polymer materials such as Polyvinyl alcohol (PVA), Poly(methyl methacrylate) (PMMA), Polypropylene (PP), Polystyrene (PS), Polydimethylsiloxane (PDMS) or photocurable resins [28,33–35]. The advantages of this approach are manifold, especially for biomedical applications. For example, the release of potentially harmful Zn^{2+} cations can be properly limited without vanishing the corresponding antimicrobial action. Moreover, a sustained release of drugs over time and a limitation of undesirable burst-release effects typically observed for pure ZnO-based systems [36] could be achieved. Finally, the addition of ZnO to the polymer matrix can result in a lower friction coefficient with respect to the pristine polymer, useful to overcome the problem of discomfort of traditional ureteral stents [37].

This work aims to propose a new composite material which combines mesoporous ZnO micropowders with a hydrogel matrix in view of fabricating smart and innovative drug-eluting ureteral stents. The composite material is designed to obtain efficient anti-inflammatory drug release, biodegradation of the device itself due to zinc cations release and thus also intrinsic antibacterial properties [30,33]. To achieve such a challenging goal, poly(2-hydroxyethyl methacrylate) (polyHEMA) or the crosslinked co-polymer poly(2-hydroxyethyl methacrylate)-co-poly(acrylic acid) (poly(HEMA-co-AA)) hydrogels were selected in view of their biodegradability and swelling properties. Mesoporous ZnO micropowders, prepared by low-cost hydrothermal route, were selected as filling agent in view of their intrinsic antimicrobial and biodegradation properties coupled with a high surface area and high level of porosity in the size range of mesopores (as defined by the International Union of Pure and Applied Chemistry, IUPAC [38]), ideal for drug uptake and retention. The composites were prepared by incorporating different percentages of the mesoporous ZnO powders into the polymeric composition. Then, two different anti-inflammatory drugs (Ibuprofen and Diclofenac, both in salts form) were loaded and the corresponding release profiles were investigated in artificial urine, even at different pH to simulate pathological conditions.

2. Materials and Methods

2.1. Materials Preparation

2.1.1. Synthesis of Flower-Like ZnO Mesoporous Microparticles

Mesoporous ZnO microparticles have been prepared using a low-cost, hydrothermal sol-gel synthesis approach. All the chemical reagents have been used as purchased and without any further purification. In total, 5.58 g of potassium hydroxide (KOH, Sigma Aldrich, Darmstadt, Germany) and 14.8 g of zinc nitrate hexahydrate ($Zn(NO_3)_2{\cdot}6H_2O$, Sigma Aldrich) were both dissolved separately in 100 mL of bidistilled water under vigorous magnetic stirring conditions. Then, the zinc nitrate solution was added dropwise to the KOH solution under vigorous magnetic stirring to allow the reaction between precursors and the formation of a sol. After gelation, the gel was placed in a closed Teflon bottle and treated at 70 °C for 4 h. After this time, a ZnO powder precipitated at the bottom of the vessel. By filtration, it was carefully separated from the solution (pH 14) and washed several times with demineralized water until pH neutralization. Finally, the ZnO powders were air dried at 60 °C overnight. At the end of the preparation, 3.0 g of ZnO powders can be produced.

2.1.2. Synthesis of Poly(2-hydroxyethyl methacrylate) and Poly(HEMA-co-AA)

In total, 3.0 g of poly(2-hydroxyethyl methacrylate, pHEMA) hydrogel have been prepared by radical polymerization method. In total, 5.0 g of 2-Hydroxyethyl Methacrylate (HEMA) and 0.1 g of initiator 2,2-Methylpropionitrile (AIBN, 98%) were mixed together in 50 mL of toluene as solvent. To avoid the inhibition of polymerization due to oxygen, the solution was first placed under continuous nitrogen flowing. Then, radical polymerization was started by activating the initiator at 70 °C under magnetic stirring for 6 h. At the end of the reaction (Scheme 1), polyHEMA is formed and toluene is removed by low-temperature distillation at 70 °C under reduced pressure conditions ($15{\cdot}10^{-3}$ bar). Poly(HEMA-co-AA) copolymer has been prepared by combining HEMA monomer solution with 5% vol. of acrylic acid (AA) monomer and ethylene glycol dimethacrylate (EGDMA) used as a crosslinking agent (Scheme 1). Further details about the preparation of the polymers can be also found in ref. [39].

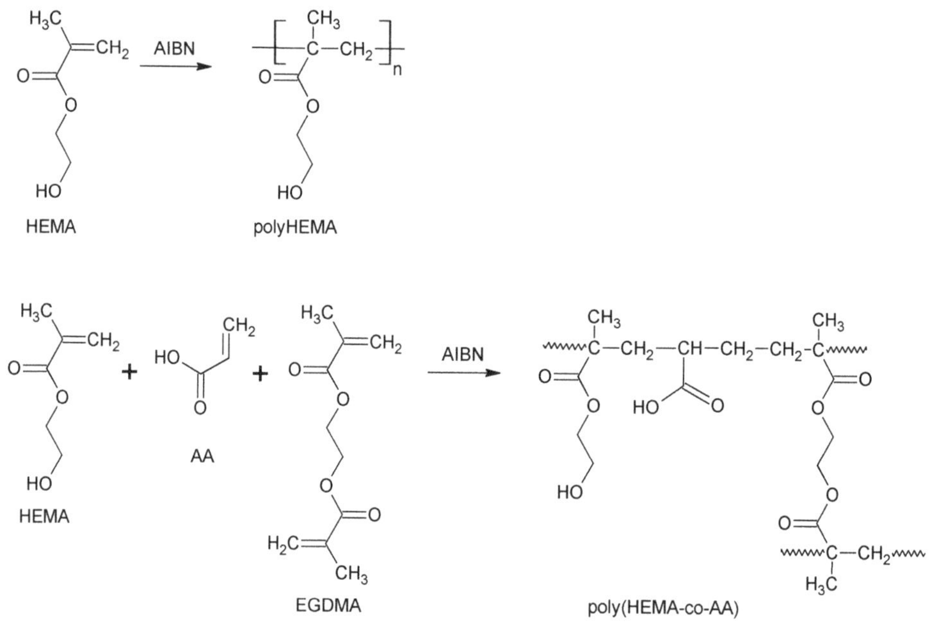

Scheme 1. The reaction scheme for the crosslinked polymers/hydrogel formation.

2.1.3. Synthesis of Composites Based on polyHEMA@ZnO and poly(HEMA-co-AA)@ZnO

The polymer@ZnO composites have been prepared by mixing the monomer solution (HEMA or HEMA-co-AA) with mesoporous ZnO micropowders in different amounts (4 mg and 40 mg), corresponding to 0.1 wt.% and 1 wt.% ZnO/polymer composite samples. In order to improve the ZnO dispersion, pre-polymerization of the solution with monomer and mesoporous ZnO powders is carried out by heating the solution in a glass tube on a hot plate at 90 °C for 30 min and until it became viscous. In total, 1% of Benzoyl Peroxide was used as an initiator (percentage relates to the total amount of monomers). The reaction is immediately interrupted by putting the beaker in ice. A rectangular silicon rubber (PDMS, polydimethylsiloxane) mold (3 cm × 2 cm × 0.3 cm in dimensions) is filled with 4 mL of the pre-polymerized solution, clamped between two Teflon plates and heated in oven for 4h at 70 °C to complete polymerization. In the specific, polyHEMA@ZnO composites have been prepared starting from HEMA monomer mixed with 1% vol. of initiator and ZnO powders. Poly(HEMA-co-AA)@ZnO were obtained from HEMA monomer, 1% vol. of AA and mesoporous ZnO powders. In the latter case, ethylene glycol dimethacrylate crosslinker (5% vol.) was also added to avoid degradation of the copolymer and improve the stability of the system. A picture representative of polyHEMA@ZnO composite sample is reported in Figure S1 of the Supporting Information.

The maximum and minimum ZnO weight percentages used in this work have been selected according to ref. [40], in order to provide the correct balance between antibacterial behavior and safety for healthy human cells of the ureter's epithelium. This will be motivated by further zinc cations release tests and the consideration reported in the Results and Discussion section.

Zinc cation release tests were performed by soaking pristine ZnO and polyHEMA@ZnO composite samples in 700 μL of Dulbecco's Modified Eagle Medium (DMEM) completed with 10% *v/v* of fetal bovine serum. The samples were prepared according to ref. [41] and incubated at 37 °C in orbital shaking conditions for different times (i.e., 2 h, 8 h, 3 days and 7 days). The free zinc cations concentration in the supernatant was evaluated through atomic absorption spectroscopy.

2.2. Ibuprofen and Diclofenac Uptake and Release

Diclofenac (DF) and Ibuprofen (IB) were selected as nonsteroidal anti-inflammatory drugs and the loading into the considered samples has been performed as follows. Both the drugs have been purchased from Sigma in sodium salt form and dissolved in water (final concentration 10 mg/mL) under vigorous magnetic stirring. The samples have been soaked into the drug solution at 37 °C to support drug solubility and under stirring conditions to prevent the formation of concentration gradients. A gas-tight cap was put on the vessel during the overall loading experiments to avoid evaporation. After 4 h, the samples have been removed and dried in an oven at 50 °C. The amount of loaded drug (11.6 ± 1.8 mg for DF and 15.3 ± 3.0 mg for IB) was estimated by weighing each dried sample before and at the end of drug uptake experiment.

The release experiments have been carried out by soaking each sample in artificial urine solution (50 mL) at physiological conditions (pH 7.4) but also at alkaline (pH 9) and acid (pH 4.5) ones to simulate pathological conditions. Artificial urine has been prepared according to ref. [42], as following: 500 mL of bidistilled water were heated at 37 °C and maintained under continuous magnetic stirring. Then, all the reagents (KCl 0.2 g/L, NaCl 8.0 g/L, Na_2HPO_4 1.14 g/L, KH_2PO_4 0.2 g/L) were added. After their complete dissolution, HCl was added until neutral pH was achieved. Then, the solution obtained was transferred in 1 L-flask and extra bidistilled water was added until reaching 1 L of artificial urine. At the end, the whole volume was transferred inside a 2L-bottle and stored in the fridge at +4 °C. When required, alkaline and acid pH conditions were obtained by adding drop-by-drop sodium hydroxide (NaOH, 1 M) and hydrogen chloride (HCl, 1 M), respectively, until reaching the desired pH value.

The samples have been soaked in orbital shaking conditions at 200 rpm (Incubating orbital shaker, Professional 3500, VWR International Ltd., Radnor, PA, USA) and maintained at a physiological temperature of 37 °C. At specific points of time (1 h, 2 h, 3 h, 4 h, 5 h, 6 h, 24 h, 48 h, 7 days, 14 days), 100 μL of the release solution was withdrawn and analyzed with ultraviolet-visible (UV-vis) spectroscopy to estimate the amount of released drug. All of the uptake/release experiments have been performed in triplicate.

2.3. Characterization Techniques

Field-emission Scanning Electron Microscope (FESEM) and Energy-dispersive X-ray (EDX) spectroscopy were carried out by using an FESEM MERLIN from ZEISS (Oberkochen, Germany). The crystalline structure was studied by X-Ray Diffraction (XRD), with a Panalytical X'Pert PRO diffractometer (Malvern Panalytical S.r.l., Milan, Italy). Cu Kα monochromatic radiation was used as the X-ray source (λ = 1.54059 Å, 40 kV, 30 mA). The diffraction patterns were collected in the 5°–80° range with a step size of 0.02° and the acquisition time was set to 100 s per step.

Nitrogen sorption isotherm (Quadrasorb SI, Quantachrome, Anton Paar QuantaTec Inc., Boynton Beach, Florida, USA) was applied to measure the porosity of the mesoporous flower-like ZnO micropowder. The Brunauer–Emmett–Teller (BET) specific surface area was applied from by multipoint method within the relative pressure range of 0.1–0.3 p/p_0. The density functional theory (DFT) model was also applied to estimate the pore size distribution of the flower-like ZnO particles.

UV-VIS absorbance spectra of drug release solutions (100 μL) were collected in the range 200–800 nm, by means of a 96-well quartz plate and a microplate reader (Multiskan™ FC Microplate Photometer, from ThermoFisher Scientific, Waltham, MA, USA). All of the UV spectra were background subtracted. Fourier Transformed infrared (FTIR) analyses were performed in attenuated total reflectance (ATR) mode with a Nicolet 5700 FTIR Spectrometer (ThermoFisher, Waltham, MA, USA). ATR-FTIR spectra acquisition was performed with 4 cm^{-1} resolution and 16 scans accumulation. Thermal Gravimetric Analysis (TGA) was performed with STA 449 Jupiter F1 (Netzsch, Selb, Germany). TGA measurements with the mass of ≈10 mg were carried out in the temperature range from 30 to 600 °C at the heating rate of 10 K/min under air atmosphere (gas flow 20 mL/min). The empty Al_2O_3 crucible was used as a reference.

3. Results and Discussion

3.1. Characterization of Flower-Like ZnO Microparticles

The morphology and the crystalline structure of the mesoporous flower-like ZnO microparticles were analyzed by means of FESEM, nitrogen sorption measurements and XRD. Figure 1a clearly highlights that the as-prepared ZnO powders consisting of spherical particles (average diameter 2.5 µm) with a flower-like morphology, due to a lateral growth of nanometer-sized crystalline facets (average thickness 15 nm). A relatively high surface area (19.6 m^2g^{-1}) and pore diameter with average size of 4 nm are evidenced from the nitrogen sorption isotherm (see Figure S2 of the Supplementary Materials.) and confirmed by the previous literature [25,43]. These pores, as well as the flower-like morphology, are responsible for the relatively high surface area and also act as preferential adsorption sites of the used drugs.

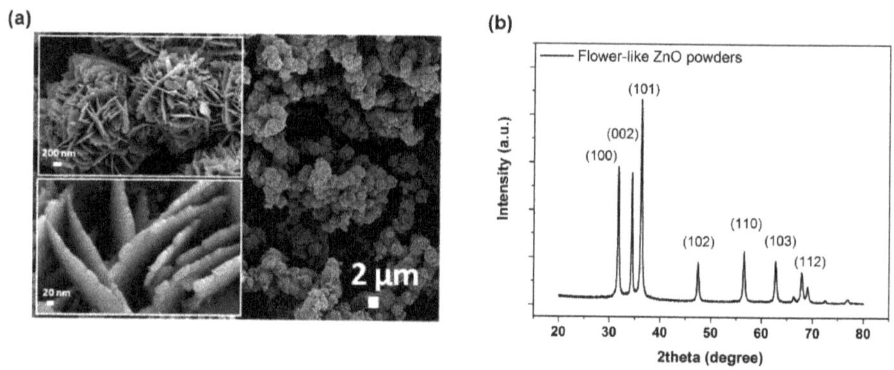

Figure 1. (**a**) FESEM images and (**b**) XRD pattern of flower-like ZnO microparticles obtained by hydrothermal method.

The XRD pattern shown in Figure 1b evidences the polycrystalline structure of the as-prepared mesoporous ZnO micropowders which crystallize in the hexagonal wurtzite structure. The diffraction pattern presents the reflection peaks typical of this crystalline phase and can be indexed according to JCPDS (file No. 80-0074) as (100) at 31.82°, (002) at 34.54°, (101) at 36.42°, (102) at 47.46°, (110) at 56.74°, (103) at 62.6° and (112) at 67.8°. The Debye–Scherrer model applied to the (101) diffraction peak leads to an average crystal size of 15 nm, which fairly corresponds to the thickness of the flower-like petals, demonstrating their nanometric nature.

The growth of this peculiar flower-like morphology is inferred to the chemical reaction used for the preparation of the ZnO powders [37]. During the hydrothermal synthesis, the coordination/electrostatic interactions allow hydroxyl groups (OH^-) to bind to Zn^{2+} metal cations. The reaction between Zn^{2+} ions and OH^- groups form the growth unit $Zn(OH)_4^{2-}$, from which ZnO can be obtained by dehydration, is as follows [28]:

$$Zn^{2+} + 2OH^- \rightarrow Zn(OH)_2 \quad (1)$$

$$Zn(OH)_2 + 2OH^- \rightarrow Zn(OH)_4^{2-} \quad (2)$$

$$Zn(OH)_4^{2-} \rightarrow 2ZnO + 2H_2O + 4OH^- \quad (3)$$

Hence, the formation of $Zn(OH)_4^{2-}$ units and the concentration of OH^- in the reaction are crucial in determining the growth rate of different crystal faces and the final ZnO morphology.

3.2. Characterization of ZnO/2-Hydroxyethyl Methacrylate-Based Composite Systems

The incorporation of flower-like ZnO microparticles within the composite material was first assessed from the morphological standpoint. Figure 2a shows the FESEM image corresponding to the polyHEMA@ZnO composite sample incorporating the lowest amount of ZnO (0.1 wt.%) used in this work. In this case, only some traces of the ZnO powders imprinted in the polymer matrix were identified and it was due to the use of a reduced quantity of ZnO. Instead, the flower-like particles were physically localized in the case of the composite containing a higher quantity of ZnO powders (1 wt.%). Figure 2b reports the corresponding FESEM image and clearly shows individual flower-like ZnO microparticles correctly incorporated inside the polymer. Similar results were obtained in the case of the copolymer (polyHEMA-co-AA), as shown in Figure S3 of the Supplementary Materials.

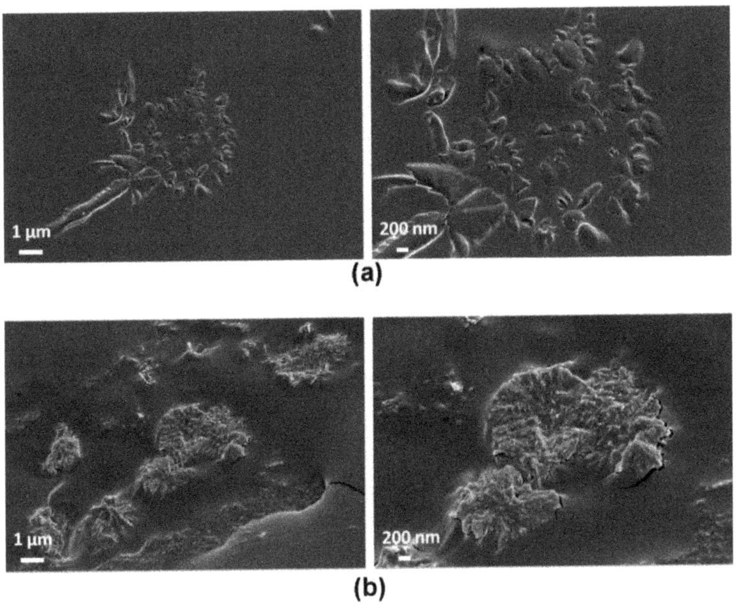

Figure 2. Morphological analysis of polyHEMA@ZnO composite samples incorporating different ZnO amounts: (**a**) ZnO_0.1 wt.%; (**b**) ZnO_1 wt.%.

The presence of ZnO powders was further confirmed by chemical composition analyses. The detection of Zn traces was confirmed by EDX analyses (Figure S4 of Supplementary Materials), with sample polyHEMA@ZnO_1% showing the higher Zn amount (0.03 at. %). The detection of the Zn element was possible on sample polyHEMA@ZnO_0.1% as well, even though the detected amount was even smaller (0.01 at.%). Even if the presence of ZnO has been only slightly detected by EDX, it is still possible to notice that polyHEMA@ZnO_1% shows a higher atomic percentage of Zn compared to polyHEMA@ZnO_0.1%. This result is thus in line with the formulation of the composites.

The actual incorporation of the ZnO powders for all the samples' typologies was further corroborated by XRD measurements, which are shown in Figure 3. The detection of the main diffraction peaks due to ZnO was observed in all the cases, independent of the polymer matrix (polyHEMA or the corresponding copolymer). Insets of each panel in Figure 3 further underline that ZnO has been correctly incorporated, as witnessed by the presence of the strong diffraction contributions coming from (100), (002) and (101) peaks due to the wurtzite ZnO phase, and positioned at 31.82°, 34.54° and 36.42°, respectively. It can be also seen that with increasing the amount of ZnO powder, the intensity of the above-mentioned peaks increases consistently, as expected. Actually, in the XRD pattern of the samples incorporating the lowest amount of ZnO (0.1 wt.%, red lines) the peaks due to ZnO are slightly noticeable. On the other hand, in the sample with 1 wt.% of ZnO (blue lines), the peaks are more clearly visible.

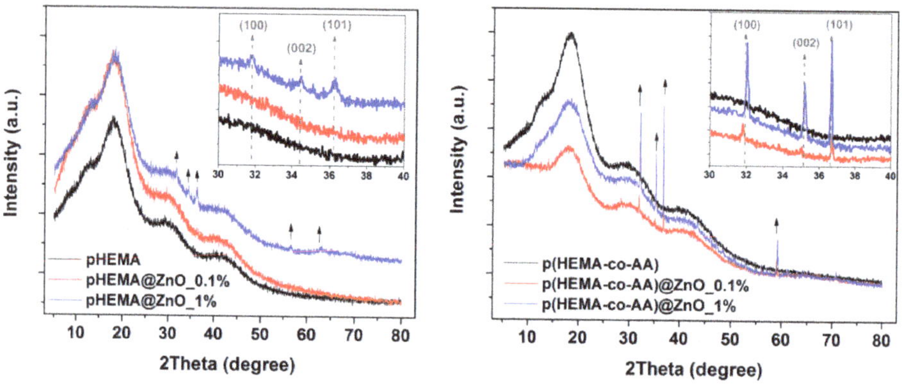

(a) (b)

Figure 3. XRD patterns of (**a**) polyHEMA@ZnO and (**b**) poly(HEMA-co-AA)@ZnO composite samples incorporating different amounts of ZnO. For each panel, arrows indicate the diffraction peaks coming from wurtzite ZnO phase. Insets show a magnification of the XRD region showing the rise of the main peaks due to (100), (002) and (101) ZnO phase.

The chemical structure of the investigated composite samples was assessed by means of FTIR spectroscopy. Figure 4a shows the FTIR spectra of the monomer used for the polymerization and of the polyHEMA hydrogel matrix, while Figure 4b shows the spectra of the corresponding composite samples including ZnO in different amounts. The only difference between the monomer and polyHEMA is the presence of H_2C=C groups in the first one and represented by two peaks at around 1600 cm^{-1} and 1000 cm^{-1}. These are more intense in the spectrum of the monomer while are slightly pronounced in the second case. This aspect allows to state that the polymerization of the starting monomer occurred correctly. However, it can be also noticed that a small amount of the monomer did not react during the polymerization, since the vibration peaks of H_2C=C groups did not completely disappear from the spectrum of polyHEMA, especially the one positioned at around 1600 cm^{-1}. Other peaks are detected as well, both in the monomer and in the polyHEMA sample. The peak over 3500 cm^{-1} relates to OH groups. The peak over 3000 cm^{-1} is associated with CH_3 and CH_2 groups. In the low wavenumber region, the peak over around 1700 cm^{-1} is associated to C=O groups while the peak over around 1250 cm^{-1} corresponds to C=C with CH_2 groups.

Concerning the FTIR spectra of the composite samples (Figure 4b), the creation of hydrogen bonding with another compound (in our case ZnO) tends to decrease the frequency of C=O stretching. The formation of carboxylic acid salts due to the surface interaction between pHEMA and ZnO is witnessed by the characteristic band at 1650 cm^{-1} due to CO_2^- asymmetric stretching vibration, which also gives rise to a band in the range 1440–1335 cm^{-1} with two characteristic peaks. Together with this C-O stretching, the O-H deformation vibration can be found in the same region, both due to the carboxylic acid of pHEMA and hydroxyl groups from the ZnO surface [44]. The presence of a broad band in the range 950–850 cm^{-1} due to Zn-OH mode is observed in the FTIR spectrum of the ZnO powders (Figure S5 of Supplementary Materials) and also in the spectra of the composite samples.

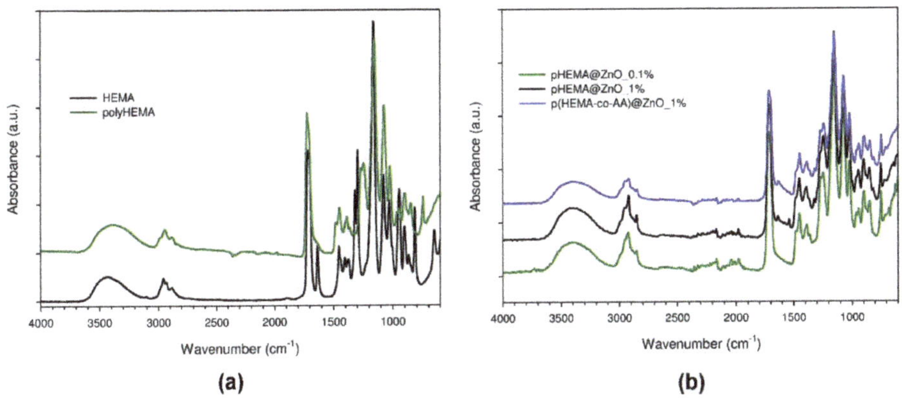

Figure 4. (**a**) FTIR spectra of HEMA monomer and of polyHEMA hydrogel obtained after radical polymerization; (**b**) Comparison of FTIR spectra of composite samples incorporating different percentages of ZnO.

TGA (Figure 5) was performed for all the samples' typologies to evaluate the thermal degradation behavior of the considered materials. Actually, an appropriate knowledge of the thermal resistance of polymer-based composites is important because of the processing conditions used. From the differential thermogravimetric (DTG) curves (Figure 5) it can be seen that the sample degradation occurs via a multistep process, as evidenced by the corresponding distinct peaks. The first step from room temperatures to around 160 °C is due to the evaporation of non-bonded water, other volatile compounds and residual unreacted monomers still present in the composites, as discussed before. The second step takes place at around 250 °C and is the most pronounced for polyHEMA with the smallest amount of ZnO. The decomposition of polyHEMA network as well as the copolymer networks start from around 200 °C. The higher thermal resistance (expressed as 2% of mass loss) for copolymer composites, with respect to the polymeric ones, is due to their crosslinked structure. The next two steps with maximum at about 330 °C and 370 °C are due to the main decomposition of polymer networks in the whole mass. The previous studies showed that depolymerization to the monomer is the dominant decomposition process for pHEMA, especially at lower temperatures [45,46]. However, the presence of ZnO powder in the polymeric samples shifts their thermal resistance towards higher temperatures.

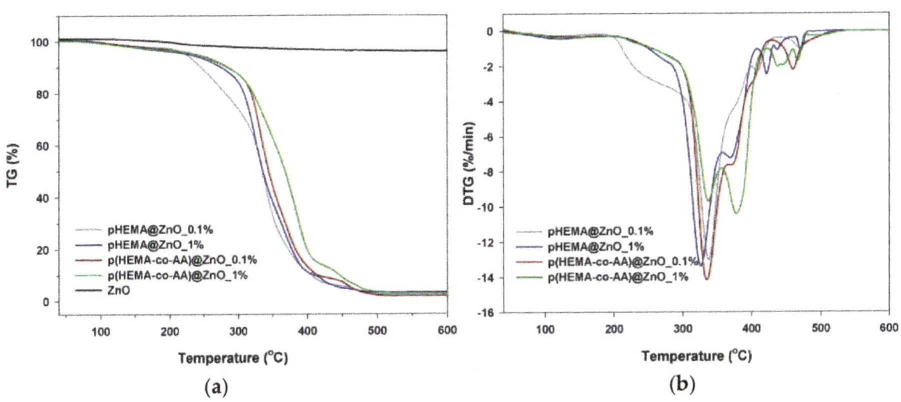

Figure 5. Thermal degradation curves of polymer-ZnO composites as well as ZnO powders as control sample (curve in black). (**a**) Thermogravimetry (TG) analysis and (**b**) Differential thermogravimetric (DTG) measurement.

To evaluate the degradation products, the emissions from TGA were evaluated and identified by FTIR analysis. Figure 6 gives a 3D spectrum of gases produced during the thermal degradation, while Figure 7 shows the FTIR spectra obtained from the 3D FTIR ones and evaluated at the maximum degradation temperature.

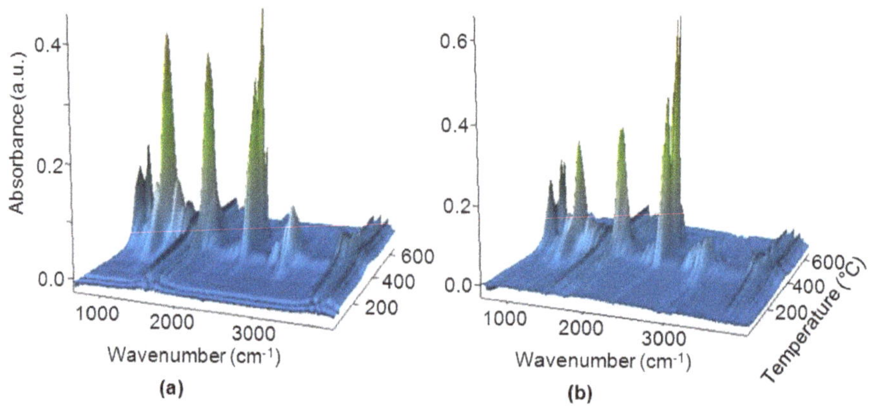

Figure 6. The 3D FTIR spectra for (**a**) polyHEMA@ZnO_0.1%; (**b**) crosslinked poly(HEMA-co-AA)@ZnO_0.1%.

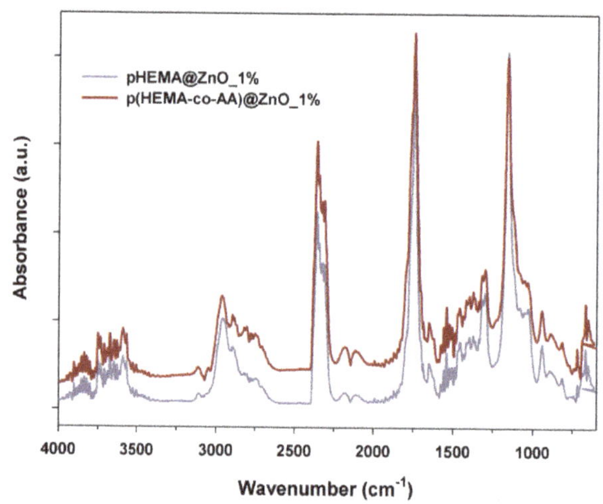

Figure 7. The FTIR spectra of the evolved gases from polyHEMA@ZnO_0.1% at 348 °C (grey) and from poly(HEMA-co-AA)@ZnO_0.1% at 348 °C (dark red).

From these results it can be stated that the degradation of polyHEMA@ZnO and poly(HEMA-co-AA)@ZnO starts from the depolymerization process to HEMA, and is followed immediately by oxidation reactions. The emission of HEMA, water and carbon monoxide and dioxide at about 350 °C, confirms such degradation patterns. However, the intensity ratio of the carbonyl band (at 1745 cm^{-1}) to carbon dioxide band (2357 cm^{-1}) indicates that the oxidation is dominant process for copolymer samples. At temperatures higher than 370 °C mainly the emission of carbon dioxide and water is observed for all samples what indicates the oxidation processes. In the spectra of Figure 7, the characteristic absorption bands for HEMA at 1745, 1638 and 1162 cm^{-1} are visible. Instead, the bands at 2357, 2311 and 670 cm^{-1} belong to carbon dioxide, while the shoulder on the absorption band at 1745 cm^{-1} is associated to another carbonyl species. Finally, the bands in the regions 3800–3600 cm^{-1} and 1600–1400 cm^{-1} are characteristic of water evaporation.

As stated in the Materials and Methods Section 2.1.3, our intention when compounding the composite materials was to provide the correct balance between antibacterial behavior and safety for healthy human cells of the ureter's epithelium. In particular, the amount of ZnO used in the composites varies between 0.1 and 1%wt/v.

ZnO at the bulk or micro-sized scale is classified as a GRAS—generally recognized as safe—substance by the US Food and Drug Administration (FDA) [47], but is also one of the most promising inorganic antimicrobial materials [48,49]. However, it can cause potential toxicity to eukaryotic cells in a dose dependent way [40,50–52]. Thus, the use of ZnO as an antimicrobial compound requires a careful balance between the material amount and the preservation of healthy living tissues and bacterial species to be killed. In a recent study performed by some of us [30], nanometer-sized ZnO were tested as nanoantibiotics against both Gram-positive and negative bacteria, such as *Escherichia coli* and *Staphylococcus aureus*, respectively. In that study, ZnO nanocrystals have shown at the highest concentration experimented, i.e., 100 μg/mL, great biocompatibility towards healthy cells (in that case pre-osteoblast cells) in terms of differentiation and proliferation of cells, and very promising antimicrobial activity against both *E. coli* and *S. aureus*.

In the present paper, a similar amount of ZnO was used in the composite formulation and allowed us to speculate that we are in the good direction to propose an efficacious biomedical device.

Additionally, a preliminary study about the dissolution behavior of ZnO microparticles in cell culture medium (to mimic the potential toxic environment against eukaryotic cells), even alone or in combination with polyHEMA, was carried out through zinc cation release tests [41], whose results are reported in Figure S6 of the Supplementary Materials. For pure ZnO microparticles, the zinc cations' concentration slowly increased with incubation time in cell culture medium. The maximum value of concentration was reached after 1 week (90.4 μg/mL), i.e., the maximum incubation time considered in this study. However, this level of released zinc cations may be cytotoxic for human cells [53]. Nevertheless, an interesting aspect arises when the ZnO microparticles are combined together with polyHEMA: a reduction of the zinc cations release was successfully obtained. Specifically, the amount of zinc cations released after incubating the samples for 3 and 7 days is reduced by 74.6% and 80.4% in comparison to that observed for the pure ZnO microparticles.

Despite using a different medium (cell culture medium instead of artificial urine solution) and a low amount of polyHEMA, it can be speculated that, in the case of the composite material discussed in this work, the use of a large amount of polyHEMA in which ZnO particles are dispersed would prevent even more the dissolution of the oxide particles, leading to the cytocompatibility of the system toward epithelial cells. An efficient zinc cation release will both allow the time-dependent biodegradation of the biomedical device and can be responsible of an interesting antimicrobial activity of the proposed composites. Further tests in this direction have to be certainly carried on in a future publication.

3.3. Drug Release in Physiological Artificial Urine

Figure 8 show the kinetic release profiles obtained for diclofenac (DF) and ibuprofen (IBU) drugs in artificial urine solution at physiological conditions (pH 7.3). All of the release profiles follow a pseudo-first-order kinetic law and the corresponding kinetic parameters (k, kinetic constant; A_{48h}, % of drug released after 48 h; A, % of drug released at the end of the experiment) are shown in Table 1. Independently of the considered material, the release kinetic was different between IBU and DF. Figure 8a highlights that DF was completely delivered after 48 h in most of the cases. A different behavior was observed only for pure polyHEMA and poly(HEMA-co-AA) samples, which showed low kinetic constants and a limited burst delivery at the same time. In particular, polyHEMA showed the lowest kinetic constant (k, 0.110 h^{-1}) and the maximum DF release (A, 96%) was reached only after 7 days (168 h). It can be also observed that the release kinetic increases according to the amount of ZnO incorporated within the polymer. Actually, both the polymers loaded with 1% ZnO are characterized by the highest kinetic constant.

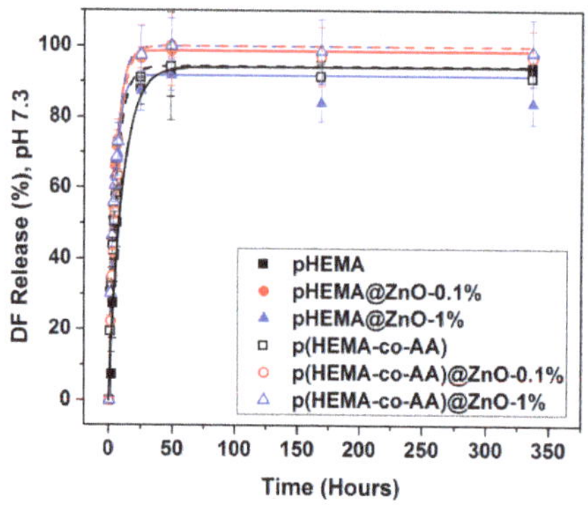

(**a**)

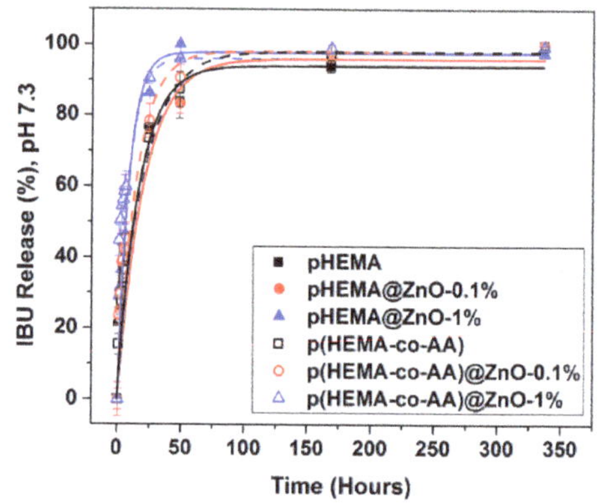

(**b**)

Figure 8. Kinetic of drug release during time, in physiological artificial urine solution: (**a**) diclofenac and (**b**) ibuprofen, both in sodium salts form. Experimental points are reported along with their error bars. All of the experimental points follow a pseudofirst-order kinetic law. Fitting curves are represented by plain lines for pHEMA-based samples and by dashed lines for p(HEMA-co-AA)-based samples.

Table 1. Kinetic parameters for DF and IBU drug release at physiological pH conditions.

pH	Composite Type		Diclofenac				Ibuprofen			
7.3	Polymer	ZnO, (w/v)	K, (h^{-1})	A_{48h}, (%)	A, (%)	R^2	K, (h^{-1})	A_{48h}, (%)	A, (%)	R^2
	polyHEMA	-	0.110	93	94	0.993	0.050	83	100	0.993
		0.1%	0.226	96	87	0.996	0.051	84	100	0.980
		1%	0.239	92	94	0.945	0.122	98	100	0.955
	polyHEMA-co-AA	-	0.178	93	94	0.997	0.050	87	100	0.992
		0.1%	0.180	100	100	0.998	0.070	91	100	0.919
		1%	0.212	100	100	0.998	0.121	96	100	0.998

On the other side, Figure 8b shows a more sustained delivery IBU over time. The release increases continuously, and it is completed after 7 days (168 h) in most of cases; only sample polyHEMA@ZnO_1% reaches the maximum release after 48 h. Concerning the release kinetic, the behavior of the investigated materials is similar to the one observed for DF: the pure polymer materials show a limited burst effect and low kinetic constants, while those incorporating the highest amount of ZnO powders show the fastest IBU release with the composite poly(HEMA-co-AA)@ZnO_1% one, having the highest release kinetic (0.120 h^{-1}).

3.4. Drug Release in Acid and Alkaline Artificial Urine

In physiological conditions, the pH value for urine is between 4.5 and 7.5, with a tendency to keep a pH value slightly acid. Figure 9a–d show the kinetic release for both DF and IBU drugs in an acidic artificial urine environment (pH 5.2). In this case, a more sustained release was observed with respect to neutral pH conditions, especially within 48 h, and it is well represented by the kinetic parameters reported in Table 2. If DF is considered, the release is strongly delayed. In particular, when a poly(HEMA-co-AA) sample is considered, DF delivery occurred only after 24 h (Figure 9b), while after 4 h for the other poly(HEMA-co-AA)@ZnO samples. In the case of polyHEMA and polyHEMA@ZnO composites, DF release started at lower release times (after 2 h). The maximum amount of drug released at the end of the experiment was generally lower than 100% and the amount of drug released after 48 h was highly lower with respect to what observed in neutral pH conditions, as discussed before. Similar aspects were found for IBU release, independently of the considered sample. The release of this drug generally started only after 2 h, and the maximum amount of drug released was lower than 100%.

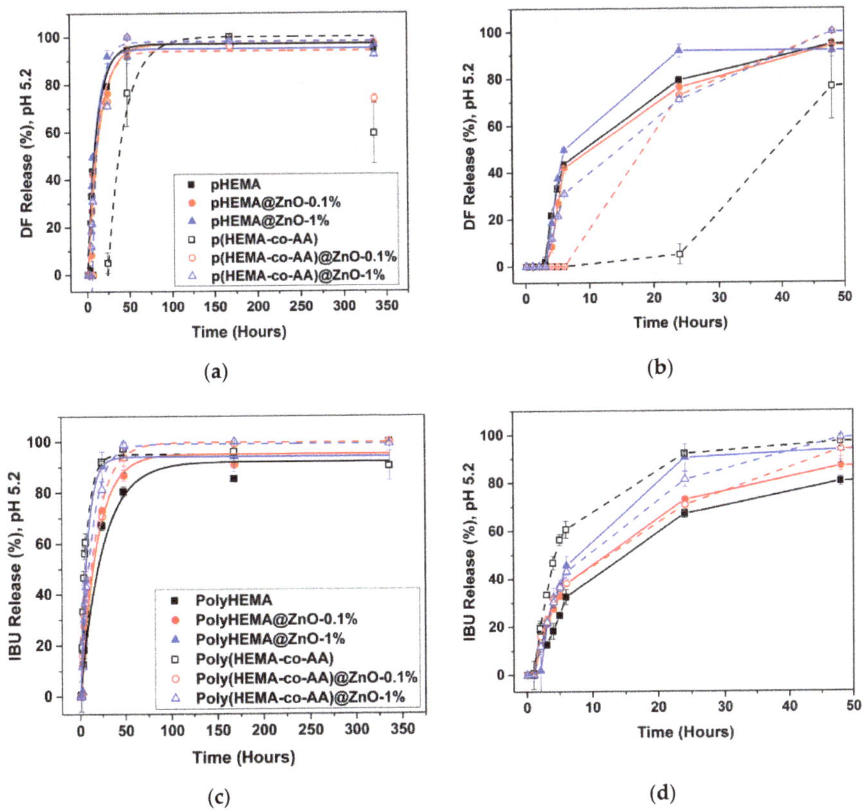

Figure 9. Drug release profiles during time in acidic artificial urine solution: (**a**,**b**) diclofenac and (**c**,**d**) ibuprofen, both in sodium salts form. Experimental points are reported along with their error bars. All of the experimental points follow a pseudofirst-order kinetic law. Fitting curves are represented by plain lines for pHEMA-based samples and by dashed lines for p(HEMA-co-AA)-based samples.

Table 2. Kinetic parameters for DF and IBU drug release at acidic pH conditions.

pH	Composite Type		Diclofenac				Ibuprofen			
5.2	Polymer	ZnO, (w/v)	K, (h^{-1})	A_{48h}, (%)	A, (%)	R^2	K, (h^{-1})	A_{48h}, (%)	A, (%)	R^2
	polyHEMA	-	0.085	95	97	0.960	0.04	80	92	0.987
		0.1%	0.070	94	94	0.953	0.055	87	95	0.993
		1%	0.100	92	95	0.945	0.130	93	94	0.955
	polyHEMA-co-AA	-	0.06	75	60	0.975	0.012	97	90	0.999
		0.1%	0.08	100	74	0.990	0.060	94	100	0.957
		1%	0.05	100	92	0.975	0.080	99	100	0.998

To simulate pathological conditions, the pH of artificial urine solution was changed to alkaline conditions (pH 9.4) and the release of DF and IBU anti-inflammatory drugs was also investigated in such situation. In alkaline pH conditions (Figure 10a,b) it can be noticed that the release increased continuously until the end of the time period, although getting slower especially within the first 48 h. Besides, all the samples reached 100% of drug release at the end of the experiments. If the amount of drug released within 48 h is considered for each sample (see Table 3) it can be noticed that in alkaline conditions, most of the samples showed a limited burst release effect with respect to neutral pH. This aspect is noticed especially in the case of IBU release.

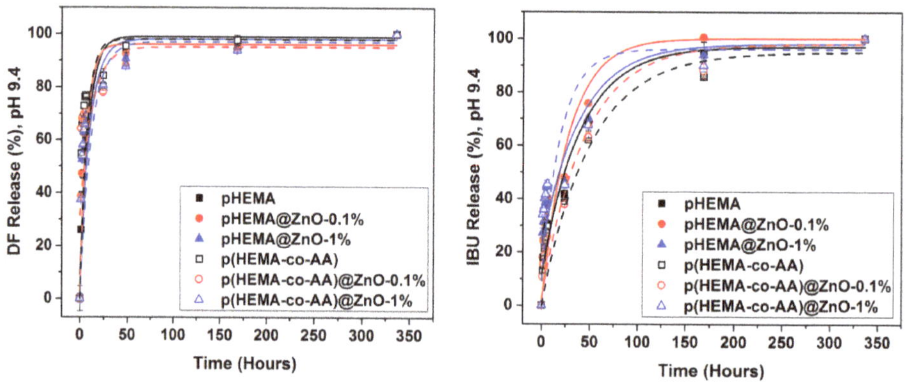

(a) (b)

Figure 10. Kinetic of drug release during time, in alkaline artificial urine solution: (**a**) diclofenac and (**b**) ibuprofen, both in sodium salts form. Experimental points are reported along with their error bars. All of the experimental points follow a pseudo first-order kinetic law. Fitting curves are represented by plain lines for polyHEMA-based samples and by dashed lines for poly(HEMA-co-AA)-based samples.

Table 3. Kinetic parameters for DF and IBU drug release at alkaline pH conditions.

pH	Composite Type		Diclofenac				Ibuprofen			
9.4	Polymer	ZnO, (w/v)	K, (h^{-1})	A_{48h}, (%)	A, (%)	R^2	K, (h^{-1})	A_{48h}, (%)	A, (%)	R^2
	polyHEMA	-	0.12	94	100	0.927	0.025	68	100	0.972
		0.1%	0.15	93	100	0.950	0.036	76	100	0.939
		1%	0.100	91	100	0.945	0.025	70	100	0.873
	polyHEMA-co-AA	-	0.15	95	100	0.978	0.02	62	100	0.899
		0.1%	0.08	88	100	0.992	0.023	63	100	0.987
		1%	0.085	88	100	0.971	0.05	68	100	0.9748

In order to explain the different release behaviors observed at different pH conditions, the chemical properties of both drug molecules should be considered. Both the DF and IBU drug used in this work are in the salt form and are characterized by an acid dissociation constant (pKa) equal to 4 at 25 °C. This quantity defines the capacity of a chemical specie to dissociate in ions upon interaction with a solution at a specific pH value. Therefore, from low pKa value it is possible to determine how strong is an acid. This dimension also depends on other physical quantities, as for instance the temperature, and in particular pKa increases with increasing of the latter. Hence, the pKa value of both drugs in salts form at 37 °C, i.e., the temperature of artificial urine solution used for release experiments, should be higher than 4. Usually, when the solution pH changes, as in this study, the balance of the dissolution reaction shifts, in accordance with the solute value of pKa. In particular, two conditions may occur, and they are reported below, where HA refers to the acid:

$$HA + H_2O \rightarrow A^- + H_3O^+$$

$$pH > pKa \rightarrow [A^-] > [HA] \rightarrow \text{dissociation of acid}$$

$$pH < pKa \rightarrow [HA] > [A^-] \rightarrow \text{acid does not dissociate.}$$

Comparing the maximum percentages of drug released in acid and alkaline pH conditions with respect to the physiological pH at 7, it can be noticed that these values are higher in case of alkaline pH and lower for acid pH. These differences occurred because in the first two studies (at pH 7 and 9.4), the solution had a pH >> pKa, which means that the drug was able to dissociate completely and to easily get free in the solution. By contrast, when the urine pH was 5, at 37 °C happened the case pH < pKa, which means that the drug dissolution was delayed and/or partially prevented.

3.5. Characterization of Post-Release Samples

After the release studies, FESEM and XRD analyses were carried out to check if some encrustations and deposition of salts occurred at the end of the immersion in AU. Besides, these analyses were taken also as a preliminary study on the degradation behavior of the investigated composite samples. Figure 11 shows the FESEM images of polyHEMA@ZnO_0.1% (panel a) and polyHEMA@ZnO_1% (panel b) acquired at the end of the studied period in AU at physiological pH conditions. The formation of encrustations and deposition of salts, mainly constituted by sodium and phosphate salts, after 14 days could be noticed from Table 4 in relatively low amount, alongside with the presence of Zn traces in the composition of the sample. These results witness that the degradation of the incorporated ZnO micropowders was not completed within two weeks of immersion in AU at physiological pH. This last aspect was further confirmed by XRD measurements (Figure 12), which indicated the presence of wurtzite ZnO powders within the composites even after the release study.

(a) (b)

Figure 11. FESEM images of salts deposits and encrustations formed at the surface of the samples after 2 weeks of immersion in artificial urine: (**a**) polyHEMA@ZnO_0.1% and (**b**) polyHEMA@ZnO_1%.

Table 4. Energy-dispersive X-ray (EDX) results of the samples polyHEMA@ZnO_0.1% and polyHEMA@ZnO_1% after 2 weeks of immersion in artificial urine.

Element	polyHEMA@ZnO_0.1%	polyHEMA@ZnO_1%
	At.%	At.%
C	16.46	14.50
O	56.09	56.86
Na	8.76	12.09
Si	0.20	0.18
P	7.80	7.52
Zn	10.68	8.85
Total:	100.00	100.00

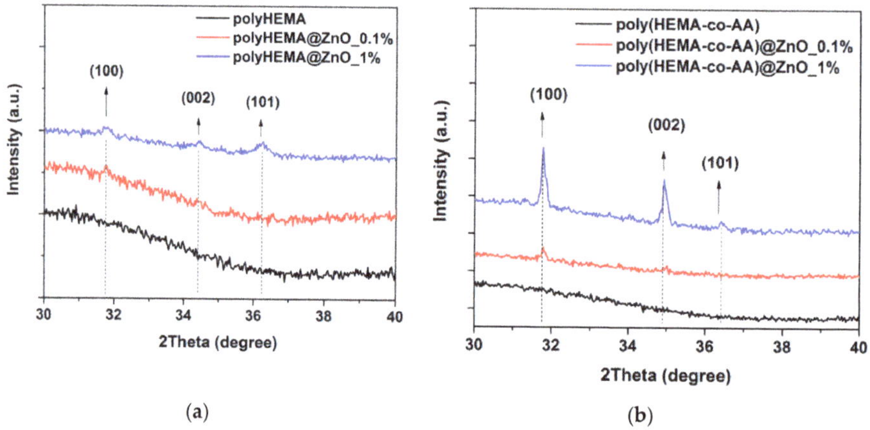

Figure 12. XRD pattern of (**a**) polyHEMA@ZnO and (**b**) poly(HEMA-co-AA)@ZnO samples at the end of the release study. Arrows indicate the diffraction peaks belonging to wurtzite ZnO phase.

The FTIR analysis of the samples after two weeks of delivery in AU solution at physiological pH showed the complete release of both drugs. Figure 13 reports the representative FTIR spectra of the copolymer composites containing 0.1% of ZnO, i.e., sample polyHEMA@ZnO_0.1%, while the FTIR spectra for the other samples are shown in Figure S7 of the Supplementary Materials. From the starting spectrum of the composite (black curve), the effective loading of both IBU and DF is confirmed by the presence of typical IR vibration modes of each drug (panel a: green curve for DF uptake and dashed green curve for pure DF drug; panel b: blue curve for IBU uptake and dashed blue curve for pure IBU). After both drug release, the typical vibration of the two drugs are no longer visible (grey spectra), and the corresponding FTIR spectra appear very similar to the one of the starting composite materials, i.e., prior to the uptake of each drug.

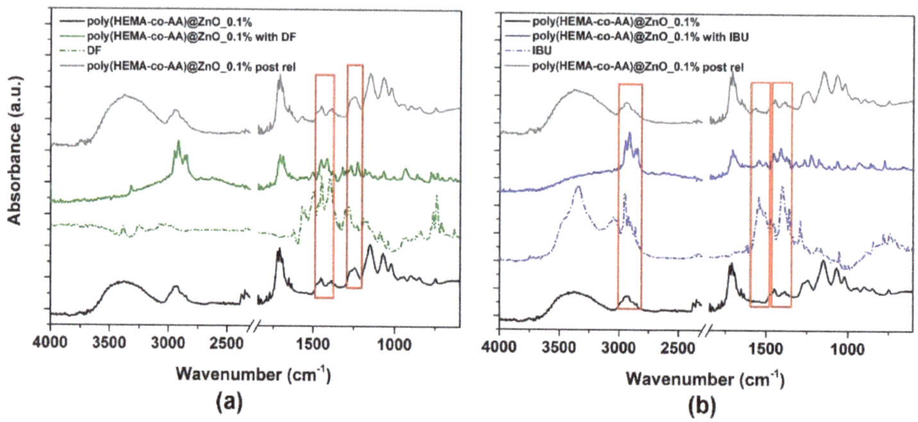

Figure 13. FTIR spectra of poly(HEMA-co-AA)@ZnO_0.1% samples in case of (**a**) Diclofenac release and (**b**) Ibuprofen release.

Both for IBU and DF release it can also be observed, from the corresponding FTIR spectra (grey curves), that no particular signatures of deposited salts are visible. As this measurement is performed on a broad size of the sample surface, it collectively shows that the presence of salts and inorganic deposit previously observed, is limited to small portions of the sample surface and hence is not so pronounced, confirming the previous results. This further supports the idea that the proposed composite materials can show high potential in fabricating drug-eluting stent combining characteristics of biocompatibility, anti-bacterial properties and prevention of encrustation deposition at least for short time periods. Certainly, more detailed studies in this direction are foreseen.

4. Conclusions

New composite materials based on the incorporation of mesoporous flower-like ZnO micropowders into polyHEMA and poly(HEMA-co-AA) hydrogels were investigated for drug eluting stents applications. Crystalline ZnO powders with flower-like morphology, good surface area and mesopore size were prepared by a low-cost hydrothermal process. Then, composite materials were fabricated combining different percentages of the inorganic ZnO powder (0.1% and 1% w/v) during the polymerization process of both polyHEMA and poly(HEMA-co-AA). The morphology analysis and structural characterization pointed out the correct integration of ZnO material inside the composite while thermal analyses demonstrated the higher stability of the copolymer, which was even more improved thanks to the addition of ZnO.

Drug release experiments in artificial urine, even performed at acidic and alkaline pH conditions, evidenced that the amount of ZnO inside the formulation did not negatively affect the ability of the considered hydrogels to store and release the anti-inflammatory drugs. On the other hand, the incorporated ZnO micropowders may work as antimicrobial agents and are expected to confer the composite additional antibacterial properties with respect to the pure hydrogel. Moreover, depending on the type of hydrogel (polymer or copolymer), the stability and release kinetics of the composite in organic fluids changed during time. At physiological conditions, a sustained drug release was observed, with the pure hydrogels showing limited burst effects and kinetic constants as low as 0.05 h^{-1}. When the pH conditions were changed from neutral to acidic/alkaline ones, the sustained release of both drugs was even more appreciable, with drug delivery starting after 4–24 h in some cases and a general reduction of the kinetic constant values, reaching the minimal values of 0.01–0.02 h^{-1} in the best cases. Furthermore, the release study pointed out that poly(HEMA-co-AA)@ZnO_0.1% is the material which has shown a good stability together with the best release trend in terms of low kinetic constant and limited burst release. Hence, this formulation satisfies the multifunctional requirements needed in the field of ureteral stent applications such as anti-bacterial effects, drug elution and biodegradability.

Supplementary Materials: The following are available online at http://www.mdpi.com/1996-1944/13/17/3821/s1, Figure S1: Picture of sample polyHEMA@ZnO_1%; Figure S2: (a) Nitrogen sorption isotherm with indication of the calculated BET surface area and (b) DFT pore size distribution of the mesoporous ZnO flower-like microparticles; Figure S3: Morphological analysis of poly(HEMA-co-AA)@ZnO composite samples incorporating different ZnO amounts: (a) ZnO_0.1 wt.%; (b) ZnO_1 wt.%; Figure S4: EDX results obtained for (a) polyHEMA@ZnO_0.1% and (b) polyHEMA@ZnO_1%; Figure S5: FTIR spectrum of mesoporous ZnO flower-like powders; Figure S6: Concentration of zinc cations released from ZnO-based samples in cell culture medium; Figure S7: FTIR spectra in case of Diclofenac and Ibuprofen release: (a,b) polyHEMA; (c,d) polyHEMA@ZnO_0.1%; (e,f) polyHEMA@ZnO_1%; (g,h) poly(HEMA-co-AA); (i,j) poly(HEMA-co-AA)@ZnO_1%.

Author Contributions: Conceptualization, M.L., M.G. and V.C.; methodology, M.L., M.G. and V.C.; software, E.D. and V.C.; validation, M.L., M.G., E.D., T.L., M.C. and V.C.; formal analysis, E.D., M.L., M.G. and V.C.; investigation, E.D., M.G., M.C., M.L. and V.C.; data curation, M.C., E.D., M.G., M.L., T.L. and V.C.; writing—original draft preparation, M.L.; writing—review and editing, M.G., T.L. and V.C.; supervision, V.C.; project administration, M.G. and V.C.; funding acquisition, E.D., M.G. and V.C. All authors have read and agreed to the published version of the manuscript.

Funding: This work was partially funded by a STSM grant from the COST Action CA16217 "ENIUS" and funded by COST (European Cooperation in Science and Technology).

Acknowledgments: The authors gratefully acknowledge Mauro Raimondo and Salvatore Guastella for their assistance with FESEM and EDX analyses.

Conflicts of Interest: The authors declare no conflict of interest.

References

1. Alnadhari, I.; Alwan, M.A.; Salah, M.A.; Ghilan, A.M. Treatment of retained encrusted ureteral Double-J stent. *Archivio Italiano di Urologia e Andrologia* **2018**, *90*, 265–269. [CrossRef] [PubMed]
2. Lange, D.; Bidnur, S.; Hoag, N.; Chew, B.H. Ureteral stent-associated complications—Where we are and where we are going. *Nat. Rev. Urol.* **2015**, *12*, 17–25. [CrossRef] [PubMed]
3. Yang, L.; Whiteside, S.; Cadieux, P.A.; Denstedt, J.D. Ureteral stent technology: Drug-eluting stents and stent coatings. *Asian J. Urol.* **2015**, *2*, 194–201. [CrossRef]
4. Staubli, S.E.; Mordasini, L.; Engeler, D.S.; Sauter, R.; Schmid, H.P.; Abt, D. Economic Aspects of Morbidity Caused by Ureteral Stents. *Urol. Int.* **2016**, *97*, 91–97. [CrossRef] [PubMed]
5. Forbes, C.; Scotland, K.B.; Lange, D.; Chew, B.H. Innovations in Ureteral Stent Technology. *Urol. Clin. N. Am.* **2019**, *46*, 245–255. [CrossRef] [PubMed]
6. Leung, J.W.; Lau, G.T.; Sung, J.J.; Costerton, J.W. Decreased bacterial adherence to silver-coated stent material: An In Vitro study. *Gastrointest. Endosc.* **1992**, *38*, 338–340. [CrossRef]
7. Cauda, V.; Chiodoni, A.; Laurenti, M.; Canavese, G.; Tommasi, T. Ureteral double-J stents performances toward encrustation after long-term indwelling in a dynamic in vitro model. *J. Biomed. Mater. Res. B Appl. Biomater.* **2017**, *105*, 2244–2253. [CrossRef]
8. Cauda, F.; Cauda, V.; Fiori, C.; Onida, B.; Garrone, E. Heparin coating on ureteral Double J stents prevents encrustations: An In Vivo case study. *J. Endourol.* **2008**, *22*, 465–472. [CrossRef]
9. Laube, N.; Kleinen, L.; Bradenahl, J.; Meissner, A. Diamond-Like Carbon Coatings on Ureteral Stents—A New Strategy for Decreasing the Formation of Crystalline Bacterial Biofilms? *J. Urol.* **2007**, *177*, 1923–1927. [CrossRef]
10. Fu, W.-J.; Wang, Z.-X.; Li, G.; Cui, F.-Z.; Zhang, Y.; Zhang, X. Comparison of a biodegradable ureteral stent versus the traditional double-J stent for the treatment of ureteral injury: An experimental study. *Biomed. Mater.* **2012**, *7*, 065002. [CrossRef]
11. Chew, B.H.; Paterson, R.F.; Clinkscales, K.W.; Levine, B.S.; Shalaby, S.W.; Lange, D. In Vivo Evaluation of the Third Generation Biodegradable Stent: A Novel Approach to Avoiding the Forgotten Stent Syndrome. *J. Urol.* **2013**, *189*, 719–725. [CrossRef] [PubMed]
12. Blum, A.P.; Kammeyer, J.K.; Rush, A.M.; Callmann, C.E.; Hahn, M.E.; Gianneschi, N.C. Stimuli-Responsive Nanomaterials for Biomedical Applications. *J. Am. Chem. Soc.* **2015**, *137*, 2140–2154. [CrossRef] [PubMed]
13. Taylor-Pashow, K.M.L.; Della Rocca, J.; Huxford, R.C.; Lin, W. Hybrid nanomaterials for biomedical applications. *Chem. Commun.* **2010**, *46*, 5832–5849. [CrossRef] [PubMed]
14. Chimene, D.; Alge, D.L.; Gaharwar, A.K. Two-Dimensional Nanomaterials for Biomedical Applications: Emerging Trends and Future Prospects. *Adv. Mater.* **2015**, *27*, 7261–7284. [CrossRef] [PubMed]
15. Zhang, Y.; Nayak, T.R.; Hong, H.; Cai, W. Biomedical applications of zinc oxide nanomaterials. *Curr. Mol. Med.* **2013**, *13*, 1633–1645. [CrossRef]
16. Laurenti, M.; Cauda, V. ZnO Nanostructures for Tissue Engineering Applications. *Nanomaterials* **2017**, *7*, 374. [CrossRef]
17. Baruah, S.; Dutta, J. Hydrothermal growth of ZnO nanostructures. *Sci. Technol. Adv. Mater.* **2009**, *10*, 013001. [CrossRef]
18. Cauda, V.; Gazia, R.; Porro, S.; Stassi, S.; Canavese, G.; Roppolo, I.; Chiolerio, A. Nanostructured ZnO materials: Synthesis, properties and applications. In *Handbook of Nanomaterial Properties*; Bhushan, B., Luo, D., Schricker, S.R., Sigmund, W., Zauscher, S., Eds.; Springer: Berlin, Germany, 2014.
19. Stassi, S.; Cauda, V.; Ottone, C.; Chiodoni, A.; Pirri, C.F.; Canavese, G. Flexible piezoelectric energy nanogenerator based on ZnO nanotubes hosted in a polycarbonate membrane. *Nano Energy* **2015**, *13*, 474–481. [CrossRef]
20. Wang, R.M.; Xing, Y.J.; Xu, J.; Yu, D.P. Fabrication and microstructure analysis on zinc oxide nanotubes. *New J. Phys.* **2003**, *5*, 115. [CrossRef]
21. Laurenti, M.; Cauda, V.; Gazia, R.; Fontana, M.; Rivera, V.F.; Bianco, S.; Canavese, G. Wettability Control on ZnO Nanowires Driven by Seed Layer Properties. *Eur. J. Inorg. Chem.* **2013**, *2013*, 2520–2527. [CrossRef]
22. Li, Y.B.; Bando, Y.; Sato, T.; Kurashima, K. ZnO nanobelts grown on Si substrate. *Appl. Phys. Lett.* **2002**, *81*, 144–146. [CrossRef]

23. Hughes, W.L.; Wang, Z.L. Controlled synthesis and manipulation of ZnO nanorings and nanobows. *Appl. Phys. Lett.* **2005**, *86*, 043106. [CrossRef]
24. Cauda, V.; Pugliese, D.; Garino, N.; Sacco, A.; Bianco, S.; Bella, F.; Lamberti, A.; Gerbaldi, C. Multi-functional energy conversion and storage electrodes using flower-like Zinc oxide nanostructures. *Energy* **2014**, *65*, 639–646. [CrossRef]
25. Pugliese, D.; Bella, F.; Cauda, V.; Lamberti, A.; Sacco, A.; Tresso, E.; Bianco, S. A Chemometric Approach for the Sensitization Procedure of ZnO Flower-like Microstructures for Dye-sensitized Solar Cells. *ACS Appl. Mater. Interfaces* **2013**, *5*, 11288–11295. [CrossRef] [PubMed]
26. Gao, P.X.; Wang, Z.L. High-Yield Synthesis of Single-Crystal Nanosprings of ZnO. *Small* **2005**, *1*, 945–949. [CrossRef]
27. Ramgir, N.S.; Late, D.J.; Bhise, A.B.; More, M.A.; Mulla, I.S.; Joag, D.S.; Vijayamohanan, K. ZnO Multipods, Submicron Wires, and Spherical Structures and Their Unique Field Emission Behavior. *J. Phys. Chem. B* **2006**, *110*, 18236–18242. [CrossRef] [PubMed]
28. Cauda, V.; Stassi, S.; Lamberti, A.; Morello, M.; Fabrizio Pirri, C.; Canavese, G. Leveraging ZnO morphologies in piezoelectric composites for mechanical energy harvesting. *Nano Energy* **2015**, *18*, 212–221. [CrossRef]
29. Garino, N.; Limongi, T.; Dumontel, B.; Canta, M.; Racca, L.; Laurenti, M.; Castellino, M.; Casu, A.; Falqui, A.; Cauda, V. A microwave-assisted synthesis of zinc oxide nanocrystals finely tuned for biological applications. *Nanomaterials* **2019**, *9*, 212. [CrossRef]
30. Garino, N.; Sanvitale, P.; Dumontel, B.; Laurenti, M.; Colilla, M.; Izquierdo-Barba, I.; Cauda, V.; Vallet-Regì, M. Zinc oxide nanocrystals as a nanoantibiotic and osteoinductive agent. *RSC Adv.* **2019**, *9*, 11312–11321. [CrossRef]
31. Xie, Y.; He, Y.; Irwin, P.L.; Jin, T.; Shi, X. Antibacterial Activity and Mechanism of Action of Zinc Oxide Nanoparticles against *Campylobacter jejuni*. *Appl. Environ. Microbiol.* **2011**, *77*, 2325–2331. [CrossRef]
32. Laurenti, M.; Lamberti, A.; Genchi, G.G.; Roppolo, I.; Canavese, G.; Vitale-Brovarone, C.; Ciofani, G.; Cauda, V. Graphene oxide finely tunes the bioactivity and drug delivery of mesoporous ZnO scaffolds. *ACS Appl. Mater. Interfaces* **2018**, *11*, 449–456. [CrossRef] [PubMed]
33. Dumontel, B.; Canta, M.; Engelke, H.; Chiodoni, A.; Racca, L.; Ancona, A.; Limongi, T.; Canavese, G.; Cauda, V. Enhanced biostability and cellular uptake of zinc oxide nanocrystals shielded with a phospholipid bilayer. *J. Mater. Chem. B* **2017**, *5*, 8799–8813. [CrossRef] [PubMed]
34. Ponnamma, D.; Cabibihan, J.-J.; Rajan, M.; Pethaiah, S.S.; Deshmukh, K.; Gogoi, J.P.; Pasha, S.K.K.; Ahamed, M.B.; Krishnegowda, J.; Chandrashekar, B.N.; et al. Synthesis, optimization and applications of ZnO/polymer nanocomposites. *Mater. Sci. Eng. C* **2019**, *98*, 1210–1240. [CrossRef]
35. Chiolerio, A.; Roppolo, I.; Cauda, V.; Crepaldi, M.; Bocchini, S.; Bejtka, K.; Verna, A.; Pirri, C.F. Ultraviolet mem-sensors: Flexible anisotropic composites featuring giant photocurrent enhancement. *Nano Res.* **2015**, *8*, 1956–1963. [CrossRef]
36. Laurenti, M.; Cauda, V. Gentamicin-Releasing Mesoporous ZnO Structures. *Materials* **2018**, *11*, 314. [CrossRef] [PubMed]
37. Zhang, Z.; Yang, M.; Yuan, J.; Guo, F.; Men, X. Friction and wear behaviors of MoS2-multi-walled-carbonnanotube hybrid reinforced polyurethane composite coating. *Friction* **2019**, *7*, 316–326. [CrossRef]
38. Rouquerol, J.; Avnir, D.; Fairbridge, C.W.; Everett, D.H.; Haynes, J.M.; Pernicone, N.; Ramsay, J.D.F.; Sing, K.S.W.; Unger, K.K. Recommendations for the characterization of porous solids (Technical Report). *Pure Appl. Chem.* **1994**, *66*, 1739–1758. [CrossRef]
39. Laurenti, M.; Grochowicz, M.; Cauda, V. Porous ZnO/2-Hydroxyethyl Methacrylate Eluting Coatings for Ureteral Stent Applications. *Coatings* **2018**, *8*, 376. [CrossRef]
40. Reddy, K.M.; Feris, K.; Bell, J.; Wingett, D.G.; Hanley, C.; Punnoose, A. Selective toxicity of zinc oxide nanoparticles to prokaryotic and eukaryotic systems. *Appl. Phys. Lett.* **2007**, *90*, 2139021–2139023. [CrossRef]
41. Carofiglio, M. *Ultrasound Stimulation of Piezoelectric ZnO Films for Cell Growth*; Politecnico di Torino: Torino, Italy, 2019.
42. Khan, L.B.; Read, H.M.; Ritchie, S.R.; Proft, T. Artificial Urine for Teaching Urinalysis Concepts and Diagnosis of Urinary Tract Infection in the Medical Microbiology Laboratory. *J. Microbiol. Biol. Educ.* **2017**, *18*. [CrossRef]

43. Lops, C.; Ancona, A.; Di Cesare, K.; Dumontel, B.; Garino, N.; Canavese, G.; Hernandez, S.; Cauda, V. Sonophotocatalytic degradation mechanisms of Rhodamine B dye via radicals generation by micro- and nano-particles of ZnO. *Appl. Catal. B Environ.* **2019**, *243*, 629–640. [CrossRef] [PubMed]
44. Socrates, G. *Infrared and Raman Characteristic Group Frequencies: Tables and Charts*; John Wiley & Sons: Hoboken, NJ, USA, 2004.
45. Mohapatra, R.; Swain, A.K.; Mohapatra, R.; Rana, P.K.; Sahoo, P.K. Poly(2-Hydroxy Ethyl Methacrylate-co-Acrylic Acid) as Novel Biodegradable Macroporous Hydrogel. *Polym. Polym. Compos.* **2005**, *13*, 807–814. [CrossRef]
46. Grochowicz, M.; Kierys, A. Thermal characterization of polymer-silica composites loaded with ibuprofen sodium salt. *J. Anal. Appl. Pyrolysis* **2015**, *114*, 91–99. [CrossRef]
47. FDA U.S. Food & Drug Administration—CFR Code of Federal Regulations Title 21. Available online: https://www.accessdata.fda.gov/scripts/cdrh/cfdocs/cfcfr/CFRSearch.cfm?fr%C2%BC182.8991 (accessed on 10 July 2020).
48. Wang, L.; Hu, C.; Shao, L. The antimicrobial activity of nanoparticles: Present situation and prospects for the future. *Int. J. Nanomed.* **2017**, *12*, 1227. [CrossRef] [PubMed]
49. Martínez-Carmona, M.; Gun'ko, Y.; Vallet-Regí, M. ZnO nanostructures for drug delivery and theranostic applications. *Nanomaterials* **2018**, *8*, 268. [CrossRef] [PubMed]
50. Zhang, Y.; Nguyen, K.C.; Lefebvre, D.E.; Shwed, P.S.; Crosthwait, J.; Bondy, G.S.; Tayabali, A.F. Critical experimental parameters related to the cytotoxicity of zinc oxide nanoparticles. *J. Nanoparticle Res.* **2014**, *16*, 2440. [CrossRef]
51. Siddiqi, K.S.; Ur Rahman, A.; Tajuddin, T.; Husen, A. Properties of Zinc Oxide Nanoparticles and Their Activity Against Microbes. *Nanoscale Res. Lett.* **2018**, *13*, 141. [CrossRef]
52. Racca, L.; Canta, M.; Dumontel, B.; Ancona, A.; Limongi, T.; Garino, N.; Laurenti, M.; Canavese, G.; Cauda, V. Zinc oxide nanostructures in biomedicine. In *Smart Nanoparticles for Biomedicine*; Elsevier: Amsterdam, The Netherlands, 2018; pp. 171–187.
53. Song, W.; Zhang, J.; Guo, J.; Zhang, J.; Ding, F.; Li, L.; Sun, Z. Role of the dissolved zinc ion and reactive oxygen species in cytotoxicity of ZnO nanoparticles. *Toxicol. Lett.* **2010**, *199*, 389–397. [CrossRef]

materials

MDPI

Communication

Self-Assembly of Asymmetrically Functionalized Titania Nanoparticles into Nanoshells

Fredric G. Svensson [1], Gulaim A. Seisenbaeva [1], Nicholas A. Kotov [1,2,3] and Vadim G. Kessler [1,*]

[1] Department of Molecular Sciences, Swedish University of Agricultural Sciences (SLU), Box 7015, 750 07 Uppsala, Sweden; fredric.svensson@slu.se (F.G.S.); gulaim.seisenbaeva@slu.se (G.A.S.); kotov@umich.edu (N.A.K.)
[2] Department of Chemical Engineering, University of Michigan, Ann Arbor, MI 48109, USA
[3] Biointerfaces Institute University of Michigan, Ann Arbor, MI 48109, USA
* Correspondence: vadim.kessler@slu.se

Received: 14 September 2020; Accepted: 27 October 2020; Published: 29 October 2020

Abstract: Titania (anatase) nanoparticles were anisotropically functionalized in water-toluene Pickering emulsions to self-assemble into nanoshells with diameters from 500 nm to 3 μm as candidates for encapsulation of drugs and other compounds. The water-phase contained a hydrophilic ligand, glucose-6-phosphate, while the toluene-phase contained a hydrophobic ligand, *n*-dodecylphosphonic acid. The addition of a dilute sodium alginate suspension that provided electrostatic charge was essential for the self-limited assembly of the nanoshells. The self-assembled spheres were characterized by scanning electron microscopy, elemental mapping, and atomic force microscopy. Drug release studies using tetracycline suggest a rapid release dominated by surface desorption.

Keywords: self-assembly; nanoshells (hollow spheres); janus particles; pickering emulsion

1. Introduction

Self-assembly of inorganic nanoparticles (NPs) has emerged as a promising pathway for engineering complex nano- and microscale structures [1,2], and understanding various natural phenomena [3]. Increasing sophisticated self-assembled nanostructured constructs, including lamellar titania sheets [4], snowflake-like structures [5], and hedgehog-particles [6] can be produced under mild conditions and *en mass*. Although it was believed before that NPs used for self-assembly are required to be homogenous in size and shape [7] to form complex objects, the latest studies indicate that polydispersed NPs can produce amazingly complex assemblies competing and/or exceeding those observed in the biological world [8]. In fact, monodispersed NPs with isotropic (i.e., spherical) morphology may not be the best 'building blocks' for engineering geometrically complex and dynamic superstructures in dispersion (rather than in solid crystals) under the conditions when a system display the thermodynamic preference to produce high-mass disorganized aggregates and has no assembly restriction related to repulsive interactions or NP symmetry. This fundamental problem can be mitigated by engineering intrinsic anisotropy in the inter-particle interactions. It can be achieved, for instance, by attaching stabilizing ligands to the particle surfaces, producing so-called Janus particles, where opposite hemispheres have opposite properties (e.g., polarity). Colloidal spheres with two or more chemical zones have attracted recent interest as anisotropic building blocks for self-assembled systems [9–12]. One way to obtain Janus particles is to partially cover a part of the particle surface and then functionalize the exposed surface with ligands to obtain an opposite polarity to the particle surface [13,14]. While the ability of NPs to act as Pickering stabilizers in an emulsion is known [15], this property has been underexploited. As the NPs migrate to the phase-boundary of two immiscible liquids to lower the interfacial energy, their surfaces will be exposed to the ligands added to the two different phases [16,17]. An area of active research involves encapsulation of biomolecules, e.g., drugs, antibodies, enzymes [18], and even living cells [19,20]. Aqueous titania sols have recently been applied to encapsulate both bacteria [21] and human immune cells with retained biological activity and good viability upon release [22]. The encapsulation of bacteria is based on deposition of particles to cover the bacterial surface, and is, therefore, not suitable for encapsulation of soluble, low-weight molecules. Rather, an emulsion route could be more efficient for capture and encapsulation of such compounds.

Titania, in both anatase and rutile phases, has shown low cell toxicity that mostly appeared to be dependent on its photocatalytic activity by production of reactive oxygen species [23]. However, titania has recently been demonstrated to induce thrombosis in contact with blood even without exposure to light [24], indicative of the incomplete understanding of biological effects of titania nanostructures. Combined with their promising chemical and biological characteristics, titania-based self-assembled superstructures can be promising for a variety of biomedical applications.

Several approaches to obtain hollow titania spheres (i.e., nanoshells) have been reported in the literature. Pang and co-workers [25] produced hollow, spherical self-assembled titania spheres by a simple hydrothermal treatment, showing promise for utilization in dye-sensitized solar cells. Seisenbaeva et al. [26] hydrolyzed molecular titanium precursors to obtain hierarchical, highly porous, spheres of titania with possible application for drug delivery. In another study, Chen and co-workers [27] utilized the Pickering emulsion effect from particles in a two-phase system. Commercial titania suspended in water was added to a n-hexadecane solution of styrene, divinylbenzene, and 2,2′-azobis (2,4-dimethyl valeronitrile). Titania nanoshells formed upon vigorous stirring and subsequently stabilized by polymerization of the organic additives at elevated temperature. A number of publications have reported self-assembled silica nanoshells, ranging from nanometer size to micrometer size [28,29]. Using the Stöber process, better particle homogeneity and size control can be achieved in the synthesis of silica NPs compared to titania NPs. This, in turn, facilitates the control over properties of the self-assembled structures and interparticle forces.

A facile method for production of anisotropically functionalized gold NPs was reported by Andala and co-workers [17]. They suspended dodecylamine-stabilized gold NPs in a water-toluene system containing hydrophobic and hydrophilic ligands with thiol functional groups. Under vigorous stirring, an emulsion formed and the Au particles acted as Pickering stabilizers by migrating to the phase-boundary between the water and toluene phases. This resulted in the formation of Janus particles. We hypothesized that the same principle could be applied to produce nanoshells from inexpensive titania nanopowders with a high degree of polydispersity taking advantage of anisotropy-guided self-assembly.

2. Materials and Methods

Disodium glucose-6-phosphate (Sigma-Aldrich Sweden AB, Stockholm, Sweden) *n*-dodecylphosphonic acid (Sigma-Aldrich Sweden AB, Stockholm, Sweden), disodium alginate (Fisher Scientific Gtf AB, Göteborg, Sweden), toluene Merck AB, Solna, Sweden), tetracycline HCl (Sigma-Aldrich Sweden AB, Stockholm, Sweden) hydrothermally synthesized titania, and deionized water (DI H_2O) were used. For characterization, a Hitachi TM-1000 (Hitachi Hightech Europe AB, Solna, Sweden) and a Hitachi FlexSEM 1000II (Hitachi Hightech Europe AB, Solna, Sweden) were used for scanning electron microscopy (SEM) imaging and elemental analyses. A Bruker atomic force microscope (AFM) FastSCAN with ScanAsyst (Blue Scientific Inc., St. John's Innovation Centre, Cambridge, UK) was used for atomic force microscopy (AFM) imaging. A Malvern Zetasizer Nano, (Malvern Panalytical Ltd, Malvern, UK) was used for dynamic light scattering (DLS). Moreover, 1 mL aqueous sample in a plastic cuvette was analyzed at 20 °C in three replicates.

A Field Electron and Ion Company (FEI) Tecnai F30 ST, (Blue Scientific Inc., St. John's Innovation Centre, Cambridge CB4 0WS, UK) with a 300 kV field emission gun was used for transmission electron microscopy (TEM). Gatan micrograph suite version 3.2 was used for analyzing the TEM data. The powder diffractogram was recorded by a Bruker D8 SMART diffractometer with APEX II charge-coupled device detector (graphite monochromator) (Blue Scientific Inc., St. John's Innovation Centre, Cambridge, UK) using k(Mo-Ka) = 0.71073 Å radiation. The hydrothermal titania was synthesized as follows: 1 mL titanium (IV) ethoxide in a Teflon container was mixed with 20 µL 0.3 mM NH_4F and, subsequently, 0.48 mL DI H_2O was added. The Teflon container was placed in a steel autoclave and treated with the following temperature program: 15 min ramp from room temperature to 50 °C (held 30 min), then ramped at 1.6 °C min^{-1} to 100 °C (held 18 h), ramped at 1.3 °C min^{-1} to 180 °C (held 24 h), followed by a natural cool down. The obtained powder was repeatedly washed and centrifuged two times with 70% ethanol (Solveco, Rosersberg, Sweden), two times with DI H_2O, and finally, two times with acetone (99%, Sigma-Aldrich Sweden AB, Stockholm, Sweden) following drying at room temperature. The titania nanopowder was characterized by powder X-ray diffraction and transmission electron microscopy.

In a typical procedure, ca. 2 mg of titania was dispersed in 2 mL of toluene with 2 mM *n*-dodecylphosphonic acid by sonication for 1 h. The glass tubes were about 2 cm in diameter and the stirring bars were 15 mm × 5 mm. Immediately after sonication, the magnetic stirring was started. Under vigorous stirring, 1 mL of 0.1 w% sodium alginate aqueous solution and 2 mL of 1.6 mM glucose-6-phosphate was added simultaneously. The stirring was continued for two hours, following removal of the stirrer and the system was left to settle. The titania usually migrated to the aqueous phase to give a white foamy appearance (Figure A1). About 20 µL was pipetted to a carbon tape coated sample holder for SEM and AFM imaging. Tetracycline was chosen as a model drug for release studies. The release was followed using a Thermo Scientific GENESYS 20 UV-vis single-beam spectrophotometer (Thermo Fisher Scientific AB, Uppsala, Sweden), measuring at 400 nm. A standard series of tetracycline dissolved in 0.9% aqueous sodium chloride was used to calculate the release through the Lambert-Beer law; $a = \varepsilon$ c l. A linear relationship (R^2 = 0.996) was obtained for the standard series.

3. Results and Discussion

As ligands for functionalization, glucose-6-phosphate (G6P, hydrophilic) and *n*-dodecylphosphonic acid (DPA, hydrophobic) were chosen. Both the phosphate and phosphonate functions have documented affinity for titania surfaces [30,31], even at lowered pH [32]. In addition, their presence is easily confirmed from phosphorous by elemental analysis compared to the carboxylate function. Titania NPs formed in the sol-gel process have size distribution in 3–5 nm window of diameters [21,26,33]. They tend to form large aggregates due to their high surface energy and sonication was used to reduce the size of the aggregates. Sonication of titania powders in pure solvents (toluene or water) apparently lead to fast re-aggregation. Sonication together with either DPA or G6P was found to markedly stabilize the particles in solution (Figure A2). Further, to stabilize the self-assembled structures, alginate, an anionic polysaccharide, was added. Alginate may interact with G6P via hydrogen bonding, and possibly also electrostatically with the titania surface, as well as increasing solution viscosity.

When increasing the amounts of titania and alginate, a competition between formation of spheres and sheets was observed (Figure A3). By vigorously stirring a small amount of titania in a system of G6P (in DI H_2O), DPA (in toluene), and dilute alginate; hollow, self-assembled titania spheres were obtained (Figure 1).

Elemental mapping confirms the presence of titanium, oxygen, and phosphorus in the spheres, Figure 2.

Carbon is present on the spheres (not shown), but in lower amounts compared to the surrounding carbon tape. The relatively high presence of phosphorous on the spheres, together with no formation of sphere in the absence of the organic ligands, indicates that they are bond to the TiO_2 NPs and are essential for self-assembly process. The spheres are dispersed on a thin layer of non-assembled materials (Figure 1a), which contributes to diffuse signals in the elemental analysis in Figure 2, particularly for phosphorus. The hollow volume of spheres, and the amount of broken/incomplete spheres, varies between batches. When dried, the spheres survive for several days on the sample holder carbon tape. This would indicate stability of the structures after presumed evaporation of encapsulated toluene. They were observed to be stable in dispersion for at least one month, as determined by SEM imaging.

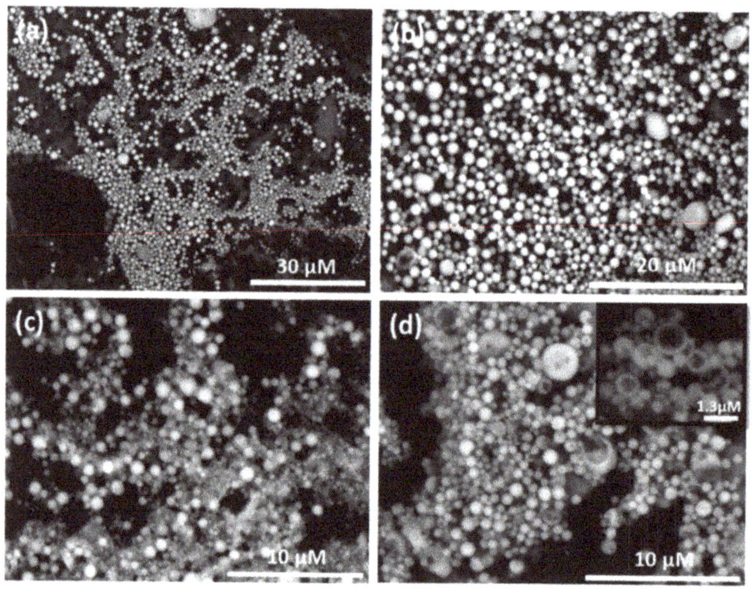

Figure 1. Scanning electron micrographs of self-assembled titania nanoshells. (**a**) Lower magnification micrograph of nanoshells dispersed over a matrice. (**b**) Micrograph of titania nanoshells of rather narrow size distribution. Average diameter calculated from 40 random nanoshells was ca. 1 µm (standard deviation (SD) 760 nm). (**c**) An example of nanoshells with higher size distribution. (**d**) Micrograph with several broken nanoshells. Average diameter calculated from 40 random nanoshells was ca. 680 nm (SD = 410 nm). The inset in (**d**) shows several incomplete nanoshells at 9000× magnification.

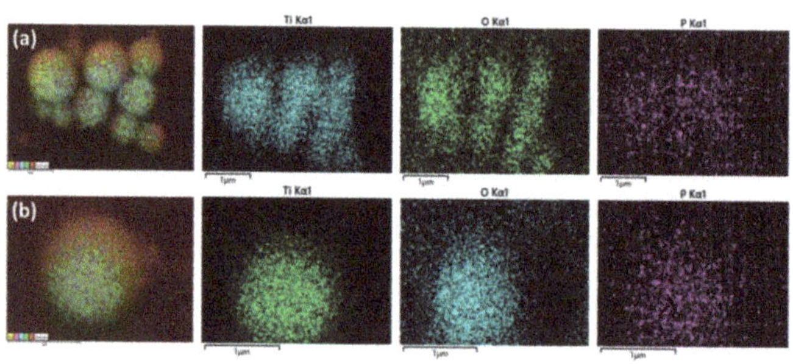

Figure 2. Elemental mapping of self-assembled nanoshells, showing the presence of titanium, oxygen, and phosphorous. Analysis of (**a**) an aggregate of self-assembled hollow spheres and (**b**) an individual hollow sphere.

The hydrothermally synthesized titania NPs was nanocrystalline in the anatase phase, as determined by powder X-ray diffraction (Figure A4). Transmission electron microscopy (TEM) micrographs revealed primary anatase particles of about 8 nm (Figure A5). Attempts with highly crystalline anatase powders (annealed >500 °C) and different titanate perovskites from solid-state syntheses failed to produce nanoshells. NP growth and increased polydispersity during high annealing temperatures is hypothesized, and have a negative impact on the self-assembly process.

To obtain more information about structural features of the surface of the spheres, atomic force microscopy (AFM) was used. It can clearly be seen that the spheres are built up by smaller aggregates and they have porous surfaces (Figure 3).

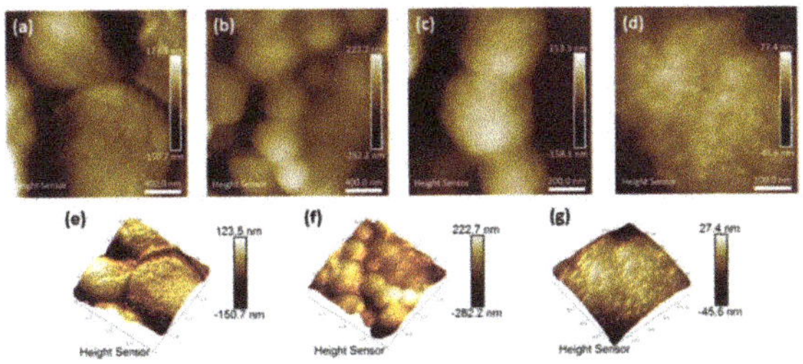

Figure 3. Atomic force micrographs of titania nanoshells. (**a**) Micrographs of a few larger nanoshells. (**b**) An aggregate of smaller nanoshells average diameter is 420 nm (SD = 80 nm). (**c**) Magnified micrograph of three small nanoshells average diameter is 416 nm (SD = 55 nm). (**d**) Magnified micrograph of the surface of a nanoshell. Nanoshell topography is shown in (**e**), and (**f**), and a three-dimensional (3D) image of a surface in (**g**).

Their spherical structure is visualized in greater detail in the AFM three-dimensional (3D) images (Figure 3e,f). The building blocks are presumably functionalized secondary particles, rather than primary NPs. The outer diameter of the nanoshells are usually within 0.5 μm to 3 μm, with an average size of approximately 1 μm, which is much smaller compared to those reported by Chen et al. [27] of about 20–50 μm. The hydrodynamic size in solution was estimated by dynamic light scattering (DLS). An average particle size of 853 nm (SD 57 nm, polydispersity index 0.451) for three replicate runs were found, which is in agreement with the observations from SEM and AFM. However, because of the wide size range (500 nm to 3 μm) of the nanoshells, the polydispersity index was relatively high. It should be noted that DLS is most accurate for particle sizes below 300 nm and the measured average should be seen as the average of a wider size distribution.

Vigorous mixing of the titania in the two-phase system results in a white, foamy aqueous phase after settling. This suggests the spheres have their outer surface coated with the hydrophilic G6P ligand, while the DPA ligands may reside on the inside. Their production without the addition of a dilute alginate suspension was not successful. The stabilizing effect may be a result of hydrogen bonding between G6P and alginate. A plausible mechanism for their self-organization is presented in Figure 4.

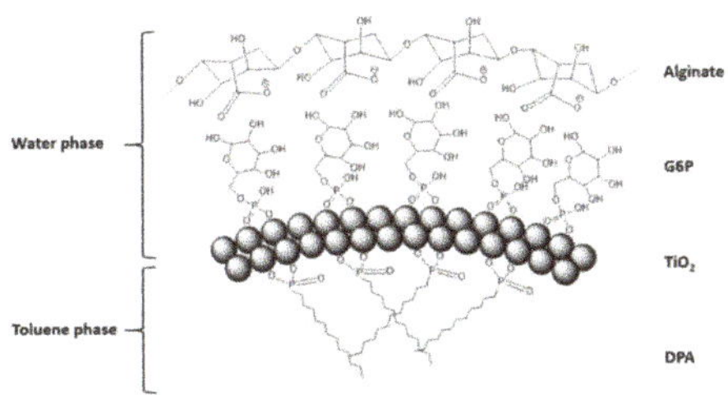

Figure 4. Proposed structure of self-assembled nanoshells. The inner side of the spheres are coated with the hydrophobic ligand (DPA), facing encapsulated toluene, and the outer side is coated with the hydrophilic ligand (G6P), facing the water phase. The alginate polymer helps stabilizing the nanoshells and may interact with the hydrophilic ligand via hydrogen bonding, and potentially via electrostatic interaction with the titania surface.

The preferential assembly of Janus NPs into nanoshells is a direct analog of self-assembly of amphiphilic surfactants in micelles because packing of the hydrophobic tails on NPs reduces their Gibbs free energy in aqueous media. The alginate macromolecules adsorbing onto the NPs and nanoshells is expected to increase the electrostatic repulsion between the constituent particles required for the self-limited self-assembly process [34,35].

4. Drug Release

Drug release from the material were investigated to identify the potential venues of the future development of the self-assembled titania nanoshells. The antibiotic tetracycline was added to the aqueous phase during mixing. It was noticed that the spherical shape turned oval when attempting to encapsulate different bioactive compounds (Figure A6). The yield of spheres also decreased, which may be explained by an emulsifying of the drugs, disturbing the interfacial tension between water and toluene. Some material was dried on glass slides and drug release was studied in 0.9% sodium chloride aqueous solution under slow magnetic stirring at room temperature. In an attempt to determine the drug loading capacity, nanoshells were dissolved in 0.1 M citric acid within 24 h. The loading capacity was estimated to ca. 1.8 mmol/g (SD 0.17·mmol/g) for tetracycline. Release was followed spectrophotometrically. A rather quick release take place during the first 15 min followed by a plateau (Figure 5).

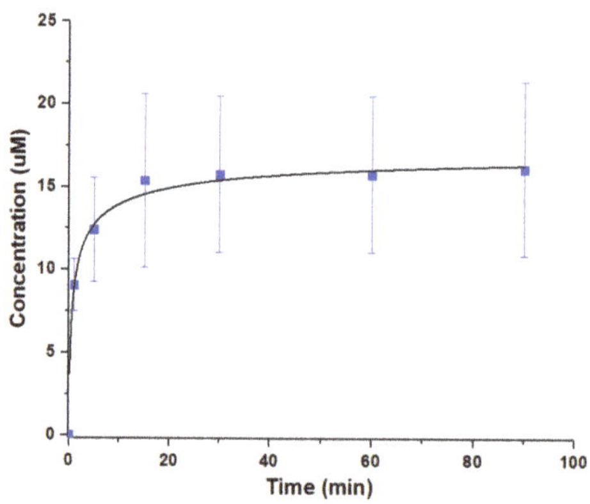

Figure 5. Release profile from the material for tetracycline in physiological sodium chloride at room temperature. Error bars are standard deviation.

This suggests release from weakly interacting surface adsorbed drug molecules accompanied by the dynamic hydrogen-bonding interactions with the alginate layer on the nanoshells. Similar rapid release profiles of weakly interacting drugs have been reported by Evdokimova et al. [36], from triclosan adsorption to pure nanocellulose, and by Kulak and co-workers [37], who studied the release of ibuprofen from porous inorganic microparticles.

5. Conclusions

Herein we report a proof of principle for the self-assembly of titania nanoshells from anisotropically functionalized titania NPs. The asymmetry of the NPs was imparted by taking advantage of their ability to stabilize Pickering emulsion step, enabling anisotropically functionalized titania NPs. Addition of alginate was found to be essential for the successful formation of the nanoshells guiding the self-assembly of polydispersed NPs away from the high-mass disorganized agglomerates toward self-limited nanoshells that displayed relatively high monodispersity. Drug release kinetics of tetracycline indicates the successful utilization of the nanoshells for drug delivery. The release profile suggest a surface adsorption of drugs with fast release.

Future work would need to be focused on increasing the nanoshell yield and their efficient separation from precursors and nanosheets. Other anisotropy-guided self-assembly could be investigated, e.g., more hydrophobic and bulkier hydrocarbon chains. Addition of bioactive compounds (which could act as emulsifiers) have an effect on the self-assembly process, and this needs optimization for potential use for encapsulation, and other applications benefiting from the simplicity and universality of the process.

Author Contributions: Conceptualization, writing—review and editing, V.G.K.; conducted overall experiments, analysis, writing—original draft preparation, F.G.S.; contributing to experiments, analysis, writing—review and editing, G.A.S.; contributing to methodology, writing—mechanism identification, review and editing, N.A.K. All authors have read and agreed to the published version of the manuscript.

Funding: This research was funded by the Swedish Research Council (Vetenskapsrådet) grant number 2014-3938. The stay of Prof. Nicholas A. Kotov at the Swedish University of Agricultural Sciences was supported by August T. Larsson Foundation and Fulbright Foundation.

Acknowledgments: The authors are grateful to Geert Cornelis (SLU) for assistance with the DLS measurements.

Conflicts of Interest: The authors declare no conflict of interest.

Appendix A

Figure A1. The material after stirring.

Figure A2. Titania sonicated in water or toluene compared with titania sonicated in the presence of ligands. The addition of ligands stabilizes the particles and less precipitation has occurred six hours after sonication.

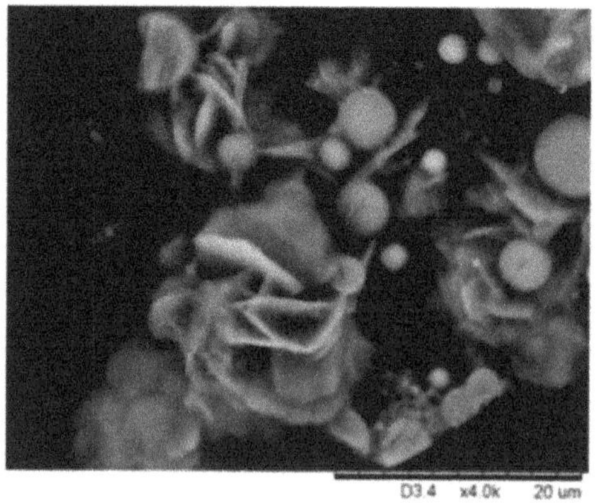

Figure A3. Competition between assembly of nanoshells and sheets when higher amounts of titania and alginate were used.

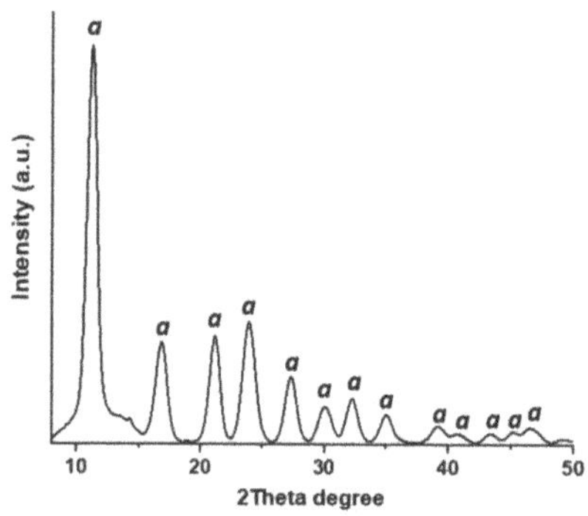

Figure A4. PXRD pattern of the hydrothermally synthesized titania. Peaks labeled *"a"* belongs to the anatase phase.

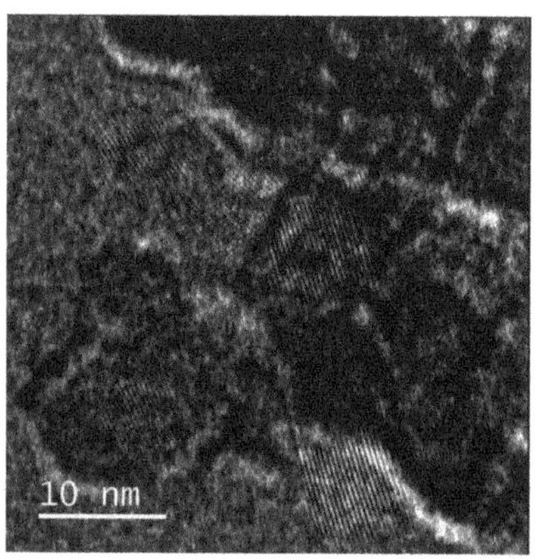

Figure A5. TEM micrograph of hydrothermally synthesized titania.

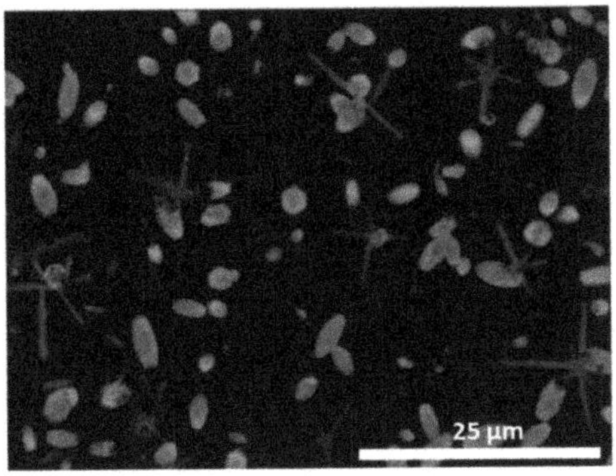

Figure A6. Change in morphology from spherical assemblies to oval assemblies at high loadings of drugs (2 mM). Star shaped assemblies are also emerging, mostly consisting of titanium, sodium and phosphorous according to EDS. These structures are presumably formed through evaporation induced self-assembly (EISA) during drying on the carbon tape (on SEM sample holder) from suspended NPs and sodium ions directed by the organic ligands.

References

1. Boles, M.A.; Engel, M.; Talapin, D.V. Self-Assembly of Colloidal Nanocrystals: From Intricate Structures to Functional Materials. *Chem. Rev.* **2016**, *116*, 11220–11289. [CrossRef] [PubMed]
2. Chen, D.; Caruso, R.A. Recent Progress in the Synthesis of Spherical Titania Nanostructures and Their Applications. *Adv. Funct. Mater.* **2013**, *23*, 1356–1374. [CrossRef]
3. Kotov, N.A. Self-Assembly of Inorganic Nanoparticles: Ab ovo(a). *Europhys. Lett.* **2017**, *119*, 66008. [CrossRef]
4. Zhu, J.; Wang, J.; Lv, F.; Xiao, S.; Nuckolls, C.; Hexing, L. Synthesis and Self-Assembly of Photonic Materials from Nanocrystalline Titania Sheets. *J. Am. Chem. Soc.* **2013**, *135*, 4719–4721. [CrossRef] [PubMed]
5. Zhang, Z.; Wang, H.; Wang, X.; Li, Y.; Song, B.; Bolarinwa, O.; Reese, R.A.; Zhang, T.; Wang, X.Q.; Cai, J.; et al. Supersnowflakes: Stepwise Self-Assembly and Dynamic Exchange of Rhombus Star-Shaped Supramolecules. *J. Am. Chem. Soc.* **2017**, *139*, 8174–8185. [CrossRef] [PubMed]
6. Deng, D.; Hao, C.; Sen, S.; Xu, C.; Král, P.; Kotov, N.A. Template-Free Hierarchical Self-Assembly of Iron Diselenide Nanoparticles into Mesoscale Hedgehogs. *J. Am. Chem. Soc.* **2017**, *139*, 16630–16639. [CrossRef]
7. Glotzer, S.C.; Solomon, M.J.; Kotov, N.A. Self-Assembly: From Nanoscale to Microscale Colloids. *AIChE J.* **2004**, *50*, 2978–2985. [CrossRef]
8. Jiang, W.; Qu, Z.B.; Kumar, P.; Vecchio, D.; Wang, Y.; Ma, Y.; Bahng, J.H.; Bernardino, K.; Gomes, W.R.; Colombari, F.M.; et al. Emergence of Complexity in Hierarchically Organized Chiral Particles. *Science* **2020**, *368*, 642–648. [CrossRef]
9. Walther, A.; Müller, A.H.E. Janus Particles: Synthesis, Self-Assembly, Physical Properties, and Applications. *Chem. Rev.* **2013**, *113*, 5194–5261. [CrossRef]
10. Su, H.; Price, C.A.H.; Jing, L.; Tian, Q.; Liu, J.; Qian, K. Janus Particles: Design, Preparation, and Biomedical Applications. *Mater. Today Biol.* **2019**, *4*, 100033. [CrossRef]
11. Chen, Q.; Bae, S.C.; Granick, S. Directed Self-Assembly of a Colloidal Kagome Lattice. *Nature* **2011**, *469*, 381–384. [CrossRef]
12. Chen, Q.; Diesel, E.; Whitmer, J.K.; Bae, S.C.; Luijten, E.; Granick, S. Triblock Colloids for Directed Self-Assembly. *J. Am. Chem. Soc.* **2011**, *133*, 7725–7727. [CrossRef] [PubMed]
13. McConnell, M.D.; Kraeutler, M.J.; Yang, S.; Composto, R.J. Patchy and Multiregion Janus Particles with Tunable Optical Properties. *Nano Lett.* **2010**, *10*, 603–609. [CrossRef] [PubMed]
14. Isojima, T.; Lattuada, M.; Vander Sande, J.B.; Hatton, T.A. Reversible Clustering of pH- and Temperature-Responsive Janus Magnetic Nanoparticles. *ASC Nano* **2008**, *2*, 1799–1806. [CrossRef] [PubMed]
15. Chevalier, Y.; Bolzinger, M.A. Emulsions Stabilized with Solid Nanoparticles: Pickering Emulsions. *Colloids Surf. A Physochemical Eng. Asp.* **2013**, *439*, 23–34. [CrossRef]

16. Harman, C.L.G.; Paterl, M.A.; Guldin, S.; Davies, G.L. Recent Developments in Pickering Emulsions for Biomedical Applications. *Curr. Opin. Colloid Interface Sci.* **2019**, *39*, 173–189. [CrossRef]
17. Andala, D.M.; Shin, S.H.R.; Lee, H.Y.; Bishop, K.J.M. Templated Synthesis of Amphiphilic Nanoparticles at the Liquid-Liquid Interface. *ACS Nano* **2012**, *6*, 1044–1050. [CrossRef]
18. Zhu, Y.; Jiang, Z.; Zhang, L.; Shi, J.; Yang, D. Sol-Gel Derived Boehmite as an Efficient and Robust Carrier for Enzyme Encapsulation. *Ind. Eng. Chem. Res.* **2012**, *51*, 255–261. [CrossRef]
19. Sakkos, J.K.; Mutlu, B.R.; Wackett, L.P.; Aksan, A. Adsorption and Biodegradation of Aromatic Chemicals by Bacteria Encapsulated in a Hydrophobic Silica Gel. *ACS Appl. Mater. Interfaces* **2017**, *9*, 26848–26858. [CrossRef]
20. Amoura, M.; Nassif, N.; Roux, C.; Livage, J.; Coradin, T. Sol-Gel Encapsulation of Cells is Not Limited to Silica: Long-Term Viability of Bacteria in Alumina Matrices. *Chem. Commun.* **2007**, *39*, 4015–4017. [CrossRef]
21. Kessler, V.G.; Seisenbaeva, G.A.; Unell, M.; Håkansson, S. Chemically Triggered Biodelivery Using Metal-Organic Sol-Gel Synthesis. *Angew. Chem. Int. Ed.* **2008**, *47*, 8506–8509. [CrossRef] [PubMed]
22. Youn, W.; Ko, E.H.; Kim, M.H.; Park, M.; Hong, D.; Seisenbaeva, G.A.; Kessler, V.G.; Choi, I.S. Cytoprotective Encapsulation of Individual Jurkat T Cells within Durable TiO_2 Shells for T-Cell Therapy. *Angew. Chem. Int. Ed.* **2017**, *56*, 10702–10706. [CrossRef] [PubMed]
23. Sayes, C.M.; Wahi, R.; Kurian, P.A.; Liu, Y.; West, J.L.; Ausman, K.D.; Warheit, D.B.; Colvin, V.L. Correlating Nanoscale Titania Structure with Toxicity: A Cytotoxicity and Inflammatory Response Study with Human Dermal Fibroblasts and Human Lung Epithelial Cells. *Toxicol. Sci.* **2006**, *92*, 174–185. [CrossRef]
24. Ekstrand-Hammarström, B.; Hong, J.; Davoodpour, P.; Sandholm, K.; Ekdahl, K.N.; Bucht, A.; Nilsson, B. TiO_2 Nanoparticles Tested in a Novel Screening Whole Human Blood Model of Toxicity Trigger Adverse Activation of the Kallikrein System at Low Concentrations. *Biomaterials* **2015**, *51*, 58–68. [CrossRef]
25. Pang, H.; Yang, H.; Guo, C.X.; Lu, J.; Li, M. Nanoparticle Self-Assembled Hollow TiO_2 Spheres with Well Matching Visible Light Scattering for High Performance Dye-Sensitized Solar Cells. *Chem. Commun.* **2012**, *48*, 8832–8834. [CrossRef]
26. Seisenbaeva, G.A.; Moloney, M.P.; Tekoriute, R.; Hardy-Dessource, A.; Nedelec, J.M.; Gun'ko, Y.K.; Kessler, V.G. Biomimetic Synthesis of Hierarchically Porous Metal Oxide Microparticles—Potential Scaffolds for Drug Delivery and Catalysis. *Langmuir* **2010**, *26*, 9809–9817. [CrossRef]
27. Chen, T.; Clover, P.J.; Bon, S.A.F. Organic-Inorganic Hybrid Hollow Spheres Prepared from TiO_2-Stabilized Pickering Emulsion Polymerization. *Adv. Mater.* **2007**, *19*, 2286–2289. [CrossRef]
28. Li, S.; Wang, F.; Dai, H.; Jiang, X.; Ye, C.; Min, J. Self-Assembly of Silica Nanoparticles into Hollow Spheres via a Microwave-Assisted Aerosol Process. *Mater. Res. Bull.* **2016**, *74*, 459–464. [CrossRef]
29. Li, M.; Zhang, C.; Yang, X.L.; Xu, H.B. Controllable Synthesis of Hollow Mesoporous Silica Nanoparticles Templated by Kinetic Self-Assembly Using a Gemini Surfactant. *RSC Adv.* **2013**, *3*, 16304. [CrossRef]
30. Nilsing, M.; Lunell, S.; Persson, P.; Ojamae, L. Phosphonic Acid Adsorption at the TiO_2 Anatase (101) Surface Investigated by Periodic Hybrid HF-DFT Computations. *Surf. Sci.* **2005**, *582*, 49–60. [CrossRef]
31. Nilsing, M.; Persson, P.; Ojamae, L. Anchor Group Influence on Molecule-Metal Oxide Interfaces: Periodic Hybrid DFT Study of Pyridine Bound to TiO_2 via Carboxylic and Phosphonic Acid. *Chem. Phys. Lett.* **2005**, *415*, 375–380. [CrossRef]
32. Svensson, F.G.; Daniel, G.; Tai, C.W.; Seisenbaeva, G.A.; Kessler, V.G. Titanium Phosphonate Oxo-Alkoxide "Clusters": Solution Stability and Facile Hydrolytic Transformation into Nano Titania. *RSC Adv.* **2020**, *10*, 6873–6883. [CrossRef]
33. Azouani, R.; Soloviev, A.; Benmami, M.; Chhor, K.; Bocquet, J.F.; Kanaev, A. Stability and Growth of Titanium-Oxo-Alkoxy TixOy(OiPr)z Clusters. *J. Phys. Chem. C* **2007**, *111*, 16243–16248. [CrossRef]
34. Xia, Y.; Nguyen, T.D.; Yang, M.; Lee, B.; Santos, A.; Podsiadlo, P.; Tang, Z.; Glotzer, S.C.; Kotov, N.A. Self-Assembly of Self-Limiting Monodisperse Supraparticles from Polydisperse Nanoparticles. *Nat. Nanotechnol.* **2012**, *7*, 479. [CrossRef] [PubMed]
35. Piccinini, E.; Pallarola, D.; Battaglini, F.; Azzaroni, O. Self-Limited Self-Assembly of Nanoparticles into Supraparticles: Towards Supramolecular Colloidal Materials by Design. *Mol. Syst. Des. Eng.* **2016**, *1*, 155–162. [CrossRef]

36. Evdokimova, O.L.; Svensson, F.G.; Agafonov, A.V.; Håkansson, S.; Seisenbaeva, G.A.; Kessler, V.G. Hybrid Drug Delivery Patches Based on Spherical Cellulose Nanocrystals and Colloid Titania-Synthesis and Antibacterial Properties. *Nanomaterials* **2018**, *8*, 228. [CrossRef]
37. Kulak, A.; Hall, S.R.; Mann, S. Single-Step Fabrication of Drug-Encapsulated Inorganic Microspheres with Complex Form by Sonication-Induced Nanoparticle Assembly. *Chem. Commun.* **2004**, *5*, 576–577. [CrossRef]

materials

MDPI

Review

Biomedical Applications of Reactive Oxygen Species Generation by Metal Nanoparticles

Roberto Canaparo [1], Federica Foglietta [1], Tania Limongi [2] and Loredana Serpe [1,*]

1 Department of Drug Science and Technology, University of Torino, Via Pietro Giuria 13, 10125 Torino, Italy; roberto.canaparo@unito.it (R.C.); federica.foglietta@unito.it (F.F.)
2 Department of Applied Science & Technology, Politecnico di Torino, Corso Duca degli Abruzzi 24, 10129 Torino, Italy; tania.limongi@polito.it
* Correspondence: loredana.serpe@unito.it; Tel.: +39-011-670-6235

Abstract: The design, synthesis and characterization of new nanomaterials represents one of the most dynamic and transversal aspects of nanotechnology applications in the biomedical field. New synthetic and engineering improvements allow the design of a wide range of biocompatible nanostructured materials (NSMs) and nanoparticles (NPs) which, with or without additional chemical and/or biomolecular surface modifications, are more frequently employed in applications for successful diagnostic, drug delivery and therapeutic procedures. Metal-based nanoparticles (MNPs) including metal NPs, metal oxide NPs, quantum dots (QDs) and magnetic NPs, thanks to their physical and chemical properties have gained much traction for their functional use in biomedicine. In this review it is highlighted how the generation of reactive oxygen species (ROS), which in many respects could be considered a negative aspect of the interaction of MNPs with biological matter, may be a surprising nanotechnology weapon. From the exchange of knowledge between branches such as materials science, nanotechnology, engineering, biochemistry and medicine, researchers and clinicians are setting and standardizing treatments by tuning ROS production to induce cancer or microbial cell death.

Keywords: metal nanoparticles; iron oxide nanoparticles; silver nanoparticles; gold nanoparticles; titanium dioxide nanoparticles; zinc nanoparticles; reactive oxygen species; photodynamic therapy; photothermal therapy; sonodynamic therapy

Citation: Canaparo, R.; Foglietta, F.; Limongi, T.; Serpe, L. Biomedical Applications of Reactive Oxygen Species Generation by Metal Nanoparticles. *Materials* **2021**, *14*, 53. https://dx.doi.org/10.3390/ma14010053

Received: 13 November 2020
Accepted: 22 December 2020
Published: 24 December 2020

1. Introduction

Nanoscience refers to the study and application of tiny materials with dimensions equal to or less than 100 nm of which many other fields, such as material science, engineering, physics, chemistry, biology and medicine, can take advantage. One of the most active areas of research in this field is the study and the development of nanostructured materials (NSMs) and nanoparticles (NPs) [1].

NSMs and NPs have unique tunable physicochemical features such as catalytic activity, electrical and thermal conductivity, light absorption and scattering that, starting from bulk counterparts, allow enhanced performance to be exploited by many different areas such as food industry, agriculture, cosmetics and, of course, medicine [2]. In the latter area, NSMs and NPs have found suitable applications in fluorescent biological labeling [3,4], pathogen detection [5], protein analysis [6], DNA structure probing [7], tissue engineering [8], separation and purification of cells and biological molecules [9], magnetic resonance imaging (MRI) contrast enhancement [10], drug and gene delivery [11,12].

Particularly, in more recent decades, NPs have been successfully used in the clinic as effective tools for alternative therapy such as photodynamic therapy (PDT) [13–15], high-intensity focused ultrasound therapy (HIFU) [16], photothermal therapy (PPT) [17] and sonodynamic therapy (SDT) [18–21]. The ever-increasing success of these therapies is due to their ability to induce the death of prokaryotic and eukaryotic cells through key cellular mechanisms such as that of induction of NP-mediated reactive oxygen species (ROS) generation [22]. Some NPs, once released into the body through different internalization methods such as oral, parenteral, inhalation administration and skin adsorption, can affect redox homeostasis both by generating ROS or lessening scavenging pathways [22,23].

1.1. ROS Generation and Oxidative Stress

Reactive oxygen species, key signaling molecules during cell signaling and homeostasis, are produced in cells by oxidases, originating from the excitation and univalent reduction of the molecular oxygen, which leads to the generation of hydroxyl radicals, superoxide anion and hydrogen peroxide [24]. Briefly, molecular oxygen generates superoxide anion, the primary ROS, via reduction of one electron catalyzed by nicotinamide adenine dinucleotide phosphate (NADPH) oxidase. Further reduction of oxygen may either lead to hydrogen peroxide or hydroxyl radicals via dismutation and metal-catalyzed Fenton reaction, respectively [25,26]. Some of the endogenous sources of ROS include mitochondrial respiration, inflammatory response, microsomes and peroxisomes. However, the occurrence of free radicals from essential byproducts of mitochondrial respiration and transition metal ion-catalyzed Fenton-type reactions mainly can regulate many signal transduction paths in a dose-dependent way. While low or medium ROS levels raise mitogenic signaling via reversible oxidations, high ROS levels lead to nucleic acids and lipid oxidation and peroxidation, resulting in cellular apoptosis and necrosis phenomena [25,27–30].

Along with free-radical and non-free radical oxygen-containing molecules, there are also reactive nitrogen, iron (Fe), copper (Cu), and sulfur species which could attribute to increased ROS formation and oxidative stress and thus impairing the redox balance [31,32]. In this regard, the appropriate physiological level of ROS is managed by antioxidant molecules such as glutathione (GSH), vitamin E, ascorbic acid, flavonoids and by detoxifying enzymes, such as catalase (CAT), glutathione peroxidase (GPX) and superoxide dismutase (SOD) [33]. According to this model, cells and tissues respond to increasing levels of oxidative stress via antioxidant enzyme systems. During conditions of mild oxidative stress, transcriptional activation of phase II antioxidant enzymes occurs via nuclear factor (erythroid- derived 2)-like 2 (Nrf2) induction. At an intermediate level, redox-sensitive mitogen-activated protein kinase (MAPK) and nuclear factor kappa-light-chain enhancer of activated B cells (NF-κB) cascades trigger a proinflammatory response. However, extremely toxic levels of oxidative stress result in mitochondrial membrane damage and electron chain dysfunction leading to cell death [24]. Therefore, perturbation of the normal redox state contributes to peroxide and free radical production that has adverse effects on cell components including proteins, lipids and DNA [34], leading to loss of cell growth, fibrosis and carcinogenesis [35–37] (Figure 1).

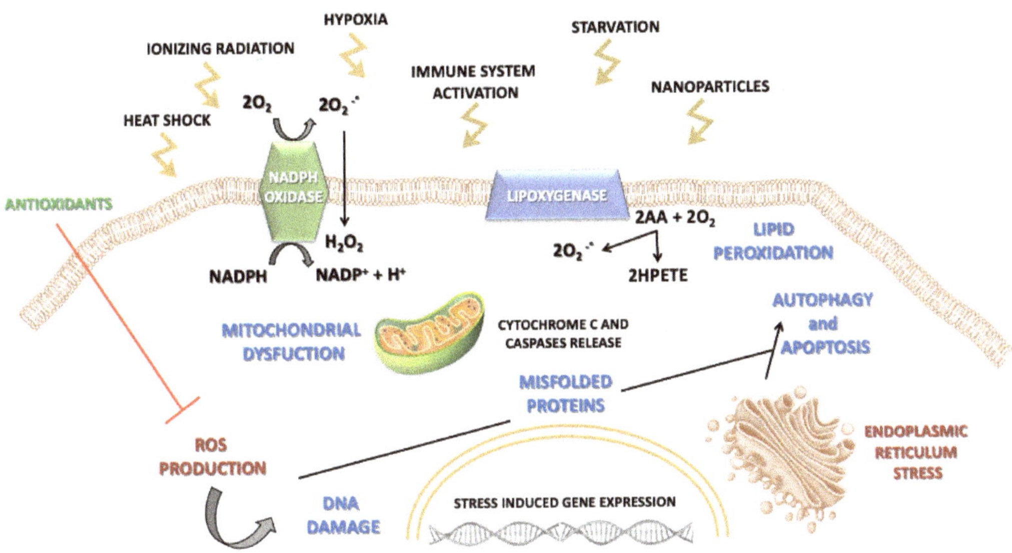

Figure 1. Schematic illustration of various triggers responsible for reactive oxygen species (ROS) generation and ROS-induced pathways leading to cell damage.

1.2. NPs-Induced Oxidative Stress

NPs of varying chemical composition such as metal oxides have been shown to induce oxidative stress and, in this regard, NPs have been reported to influence intracellular calcium concentrations, activate transcription factors and modulate cytokine production via generation of free radicals [22,25,38–41]. The main key factors involved in NP-induced ROS include: prooxidant functional groups on the reactive surface of NPs, active redox cycling on the surface of NPs due to transition metal-based NPs (MNPs), and particle-cell interactions. With regards to these key factors, several studies have shown the significance of reactive particle surfaces in ROS generation [25,39,42,43].

Free radicals are generated from the surface of NPs when both the oxidants and free radicals bind to the particle surface. Moreover, reduced particle size results in structural defects and can alter electronic properties on the NP surface, thereby creating reactive groups [44,45]. Within these reactive sites, the electron donor or acceptor interact with molecular oxygen to form superoxide anion which in turn can generate additional ROS via Fenton-type reactions [46]. For instance, NPs such as silica (Si) and zinc (Zn) with identical particle size and shape lead to diverse cytotoxicity responses due to their surface properties. Zinc oxide (ZnO), being more chemically active than silicon dioxide (SiO_2), leads to increased superoxide anion formation, resulting in oxidative stress [47]. Moreover, the mechanism for NP-mediated ROS generation can be influenced by physicochemical features of NPs such as size, chemical structure, surface area and charge. Furthermore, transition metals such as Si, Zn, Cu, Fe, chromium (Cr) and vanadium (V) are associated with ROS generation through Fenton and Haber-Weiss reaction mechanisms [42]. In Fenton responses, a transition metal ion, reacting with hydrogen peroxide, yields hydroxyl radicals and an oxidized metal ion [26]. Metal-based NPs, such as Cu and Fe, affect oxidative stress by way of Fenton reactions. On the other hand, the Haber-Weiss reaction explains the generation of hydroxyl radicals via a reaction between hydrogen peroxide and oxidized metal ions [26,37,45]. Furthermore, cobalt (Co), Cr and V NPs can catalyze both Haber-Weiss and Fenton responses, considering that the Fenton reactions are also implicated in iron-oxide NPs (IONPs)-induced ROS generation processes [31]. Finally, some NPs promote the activation of intercellular radical-inducing systems such as the MAPK and NF-κB pathways [48].

In addition to the prooxidant effect of NPs, ROS are also induced endogenously where the mitochondrion is a major cell target for NP-induced oxidative stress. Specifically, once NPs gain access into the mitochondria, they stimulate ROS via impaired electron transport chain, structural damage, activation of NADPH-like enzyme systems and depolarization of the mitochondrial membrane [49,50].

1.3. *NP-Induced Cell Death*

Apoptosis has been implicated as a major mechanism of cell death caused by NP-induced oxidative stress [51,52]. Among the different apoptotic pathways, the intrinsic mitochondrial apoptotic pathway plays a major role in metal oxide NP-induced cell death, since mitochondria are one of the major target organelles for NP-induced oxidative stress [50]. High levels of ROS in the mitochondria can result in membrane phospholipid damage and in mitochondrial membrane depolarization [53]. A small proportion of electrons escapes the mitochondrial chain and interacts with molecular oxygen to form superoxide anion which later gives rise to hydrogen peroxide or partially reduces to damaging hydroxyl radicals. NPs can catalyze the superoxide anion generation either by blocking the electron transport chain or accelerating electron transfer to molecular oxygen [54,55]. Various metal oxide NPs including Zn, Cu, titanium (Ti), and Si elicit ROS-mediated cell death via mitochondrial dysfunction [56–58].

1.4. *Introduction to Metal-Based NPs*

Metal-based NPs have been used to revolutionize several fields including sensors, catalysis, optoelectronic materials and biomedical science. Such widespread applications are attributable to their electrochemical and physical properties, reflecting their small sizes and reactive surfaces. Their fixed particle mass, high aspect ratio and particle surface bioreactivity tailor them to meet the needs of specific applications. However, a high surface-to-volume ratio makes MNPs extremely reactive, particularly with regards to free radical generation [59,60]. Furthermore, nanoscale dimensions enhance cellular uptake and interaction with biological tissues. Metal-based NPs can generate free radicals via Fenton-type reactions that react with cellular macromolecules and induce oxidative stress [61].

In this review, the authors mainly discuss the role of MNPs in ROS generation for biomedical applications with special emphasis on highly selective approaches such as photodynamic, photothermal and sonodynamic therapy.

2. Metal-Based Nanoparticle Classes and Their Biomedical Applications

Among NPs we can identify two main groups: (i) organic NPs including liposomes, polymeric NPs, carbon-based NPs and dendrimers; and (ii) inorganic NPs including QDs, metal oxides and metal and magnetic NPs [62]. Referring to the metallic ones, it must be said that they can be designed and produced through different methods of synthesis and functionalization (Table 1). To improve their biotechnological drug delivery and theranostic applications, surface functionalization can be achieved with surfactants, polymers, drugs, oligonucleotides, peptides or antibodies.

A wide range of MNPs such as silver (AgNPs), gold (AuNPs), IONPs, zinc oxide (ZnONPs) and titanium dioxide (TiO_2NPs) NPs have been mainly exploited after a whole series of optimizations and customizations for improving their applicability as therapeutic and/or diagnostic agents.

Table 1. Main advantages and disadvantages of metal-based nanoparticles (NPs).

	Advantages	Disadvantages
Metal-Based NPs	Biocompatibility High oxidation efficacy High photostability High binding affinity Low cost Surface enhanced Raman scattering Strong plasma absorption Biological system imaging Determine chemical information on metallic nanoscale substrate	Instability Impurities loaded during their synthesis Difficulty in synthesis Thermal decomposition

2.1. Iron-Oxide Magnetic NPs

In biomedical applications, magnetic nanoparticles are characterized by small inorganic crystals (<20 nm diameter) of magnetic material that can result in a core-shell configuration by their coating with an organic layer or in a multicore-shell configuration by their embedding in an organic matrix [63]. The inorganic crystals of interest can consist of several materials with ferromagnetic or superparamagnetic behaviors (Figure 2). However, the so-called iron oxides such as magnetite and maghemite are the most prevalent materials [64]. In this case, the small size of the inorganic crystal enables the formation of particles with a single magnetic domain characterized by superparamagnetic properties. Moreover, as previously described, an organic shell is usually added to the magnetic NPs in order to give them colloidal stability in biological and aqueous fluids [64,65].

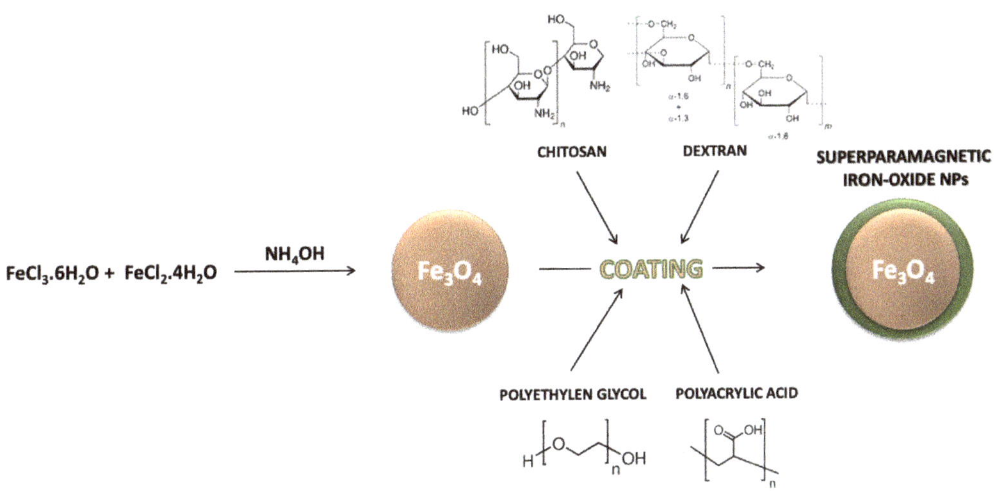

Figure 2. Iron NPs (Fe_3O_4 NPs) can be prepared by the chemical coprecipitation method and enriched with an external coating, including polyethylen glycol, polyacrylic acid, chitosan or dextran.

The ability of magnetic NPs to react with applied magnetic fields by translation (in magnetic field gradients), physical particle (in alternating and rotating fields) or internal dipole rotation (in alternating and rotating magnetic fields) shows great interests for biomedical applications [63]. The magnetic field energy can be locally converted into either thermal energy or mechanical forces and torques. Magnetic NPs are unique in possessing abilities that allow the external control of their movement and the use of their mechanical forces/torques on biological structures.

Thanks to the magnetic NP properties mentioned before, iron oxide NPs are studied for a broad range of biomedical applications such as MRI contrast agents [66], magnetically targeted drug delivery [67], magnetically assisted gene transfection [68], magneto-mechanical actuators of cell surface receptors [69], magnetically triggered drug release [70] and magnetic fluid hyperthermia (MFH). In this latter biomedical application, magnetic NPs are placed in contact with cancer tissues after which an alternating magnetic field (AMF) is employed leading to heat dissipation until an adequate thermal dose provokes cell death by different mechanisms [71]. Recently, some authors have demonstrated that one of those mechanisms refers to the generation of ROS [72,73]. Therefore, one potential cancer cell death mechanism might be the ROS production from iron-oxide NPs in combination with AMF. This phenomenon can be related to enhanced kinetic activity of the Fenton-like reaction or to the decreased ability of a cancer cell to scavenge ROS when high temperatures are present in the system [73]. For this reason, hyperthermia as a biomedical application has gained a lot of attention with more than 350 ongoing clinical trials using magnetic NPs and hyperthermia in the USA and Europe [63]. Furthermore, hyperthermia has been also used as an adjuvant treatment to improve anticancer treatments like chemotherapy and radiotherapy and indicates that MFH has a promising role as an adjuvant to chemotherapy by potentiating the effects of anticancer agents.

Interestingly, a recent in vitro study described a sonodynamic-mediated effect by employing low-intensity ultrasound (US) in combination with iron-oxide NPs and demonstrated increased production of ROS. Indeed, it is believed that US exposure, such as that in AMF, facilitates the iron release necessary to trigger the Fenton reaction which ultimately is responsible for generation of ROS [74]. The authors used US with 1 MHz intensity and different iron-oxide NPs concentrations as the sonosensitizer to explore their combined activity on the breast cancer cell line MCF-7. They considered four different groups: cells without any treatment, cells treated with iron-oxide NPs, cells exposed to US and cells treated with the sonodynamic combination of iron-oxide NPs and US. A significant decrease in cell proliferation was observed when MCF-7 cells, in in vitro experiments, underwent the combined treatment with US (at a frequency of 1 MHz with a 5 cm^2 probe, exposing cells for 1 min to a horizontal beam of continuous US wave at fixed output intensities of 2 W/cm^2) and iron-oxide NPs compared to the control group, iron-oxide NPs group and US group, suggesting that the sonodynamic effect of US and iron-oxide NPs might be due to ROS generation.

2.2. *Silver NPs*

Among MNPs, silver NPs (AgNPs), ranging between 1 and 100 nm in size, are very attractive due to their remarkable optical, thermal conductivity and electrical properties [46], which support their main role in the industrial applications of photonics, microelectronics and catalysis. In particular, AgNPs have great potential in a broad range of nanomedicine applications as biomedical device coatings, antimicrobial agents, imaging probes, drug-delivery carriers and diagnostic and optoelectronic platforms [75].

Focusing our attention on their antimicrobial activity, AgNP-cytotoxicity is characterized by their ability to release silver ions from their surface when placed in contact with an aqueous environment, as the particle surface dissolves. The release of Ag^+ ions is affected by several factors including the size and shape of NPs, capping agent and colloidal state. The interaction of silver ions with thiol groups on bacterial cell surface is caused by the large number of sulfur-containing proteins. Therefore, AgNPs can influence bacterial cell viability by their interaction with sulfur-containing proteins into the bacteria cell membrane [76]. Ag^+ ions work by substituting other essential metal ions, such as Zn^{2+} and Ca^{2+}, in critical bacterial enzymes and proteins, provoking damage of cellular respiration and cell death. AgNPs can also anchor to the surface of the bacterial cell wall and penetrate it, causing structural changes to the membrane or increasing its permeability. All of these phenomena lead to cell death. Alongside this, it has been proposed that silver ions, particularly Ag^+, released from AgNPs can provoke an interaction with DNA phosphorus moieties, causing the inactivation of DNA replication [76]. Furthermore, AgNPs can react with the sulfhydryl groups of metabolic enzymes that belong to the bacterial transport chain of electrons, proving their inactivation.

ROS and free-radical generation are another AgNP mechanism causing a cell-death process, demonstrated in in vitro studies. Indeed, the potent cytotoxic activity of AgNPs and ability to sustain antibacterial, antifungal and antiviral activity and, for some authors, also their anticancer activity [77], is mainly due to their ability to induce ROS and free-radical species production, such as hydrogen peroxide, superoxide anion, hypochlorous acid, hydroxyl radicals and singlet oxygen [78,79]. Regarding their antibacterial role, once the AgNPs are in contact with bacteria, the free radical-mediated pore generation into the cell wall is the leading mechanism of bacterial cell death. Therefore, some authors have investigated how to improve the AgNPs' antimicrobial activity via localized surface plasmon resonance (LSPR) to increase the ROS generation by exposure to visible light [80].

The LSRP is related to the electron collective oscillation at the metallic structure interface, which can be induced by the electron-magnetic interaction of the metal with light at an appropriate wavelength. In other words, LSPR is an optical phenomenon generated by conductive nanoparticles, smaller than the incident wavelength, interacting with light [81]. In this regard, Vasil'kov and colleagues have published research where they have investigated how antibacterial properties of AgNPs increased when the plasmon resonance effect occurs due to a 470 nm laser radiation application for 5 min, with a power of 5 mW, in a Petri dish containing AgNPs, Gram-positive and Gram-negative bacteria [82]. The main objective of this study was to increase the antibacterial activity of AgNPs by exploiting the capability of LSPR to influence the electronic state of the particles and increase the number of silver ions (Ag^+) which ultimately are responsible for the antimicrobial effect. Recently, research from da Silva and colleagues [80] has shown a link among AgNPs, LSPR, ROS and the AgNPs' antimicrobial activity. The authors claim that, even if some studies support the idea that Ag^+ plays a main role in antibacterial activity [83], together with the capability of AgNPs to introduce nicks in the cytoplasmic membrane [84], AgNPs and Ag^+ can also induce ROS formation, which damages the cell cytoskeleton, oxidizing nucleic acids and proteins, leading to potential chromosomal aberrations and cell death [85]. In this study, the peculiar association between the bactericidal effect of AgNPs, along with an intracellular ROS increase, probably as generated by LSPR via use of a LED floodlight (50 W) on the AgNPs' surface, was explained. This finding, in our opinion, seems very promising since it may broaden the opportunity, as we have mentioned before, to also study the cytotoxicity of AgNPs against cancer cells [86,87].

2.3. Gold NPs

Another noble metal with great abilities and promising results in medicine is gold (Au) [88]. In particular, gold nanoparticles (AuNPs) have demonstrated anticancer properties derived from different mechanisms which can be explained by varied Au properties. AuNPs can be exploited for anticancer purposes via several approaches, such as drug-delivery, anti-angiogenic, photothermal and photodynamic effects [89].

The AuNPs' photothermal application is due to the multiplicative effects of increased local absorption of laser radiation at near-infrared (NIR) frequencies by LSPR, inducing hyperthermia in cancer tissue, revolutionizing the traditional and widespread laser hyperthermia of tissues [90]. Additionally, visible light irradiation can allow for hyperthermia by LSPR, as recently demonstrated by Mendes et al., in which 14 nm AuNPs were combined with green laser light despite the fact that therapeutic efficacy of such an approach is limited in cancer due to poor penetration of light through tissue [91,92].

In this regard, to overcome the low light penetration into tissue, it has also been proved that Au-based nanotherapeutics can absorb radiofrequency (RF) and produce heat, giving the possibility to treat deeply localized tumors by using Au and hyperthermia-based options [93]. More intriguing is the approach shown by Brazzale and colleagues, where targeted AuNPs might be activated and kill cancer cells by US [19] (Figure 3).

In this pioneering in vitro work, the role of targeted AuNps as sonosensitizers in SDT, an innovative anticancer approach where, it is generally accepted, a non-toxic molecule or system (chemical actuator), i.e., the sonosensitizer, is activated by US (physical activator), yielding oxidative damage by ROS generation, and consequent cancer cell death [94,95]. More specifically, the authors suggest that US, used at the frequency of 1.866 MHz for a total 5 min of exposure, through a physical phenomenon called sonoluminescence [96], might drive the AuNPs' plasmonic effect, as derived from LSPR, to be able to convert the photon energy to heat in order to induce cellular damage via ROS production. Analysis of intracellular ROS production was investigated in two different cell lines, HCT-116 and KB, and demonstrated that, while cells incubated with AuNPs in the absence of US and US alone did not encounter an increase in intracellular ROS productions, cells that were exposed to both AuNPs and US, i.e., sonodynamic treatment, were subjected to a significant increase in ROS production and therefore enhanced death rates.

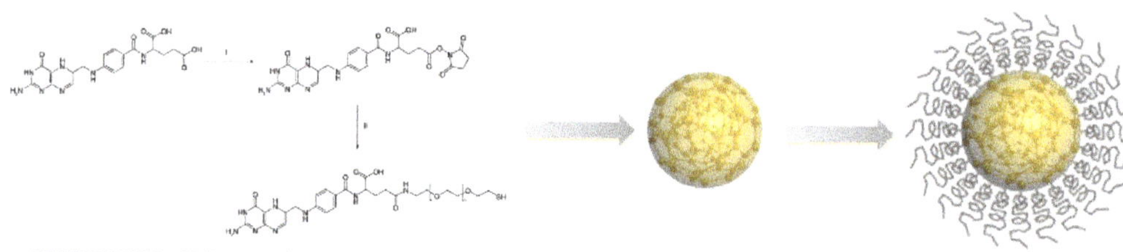

Figure 3. Gold nanoparticles (Au-NPs) can be decorated by adding the folic acid carboxyl group. Folic acid was activated by carboxyl groups by N-hydroxysuccinimide (NHS) and dicyclohexylcarbodiimide in anhydrous dimethyl sulfoxide (DMSO) and conjugation of NHS-ester activated folate to NH_2-PEG3.5kDa-SH (FA-PEG-SH) in anhydrous DMSO in presence of triethylamine [19].

2.4. Titanium Dioxide NPs

Titanium dioxide (TiO_2) is one of the most extensively used nanomaterials for several applications [97], but the photocatalytic properties of TiO_2NPs have raised many issues as a result of ROS generation while UV irradiation is performed. Indeed, electrons in the TiO_2 valence band absorb the photon energy under UVA irradiation, and jump to the conduction band, allowing extraction of electrons from water or hydroxyl ions generating hydroxyl radicals by valence band holes. Other methods of ROS formation, such as superoxide anion and singlet oxygen by additionally mechanisms, have also been demonstrated [98,99]. However, the photocatalytic properties of this NP make TiO_2 a valuable competitor for some biomedical applications, such as in killing microorganisms [100,101] and treating malignant tumors. The latter application has been investigated since 1992, when Cai et al. studied the effect of photoexcited TiO_2 on cancer cells in in vitro studies [102]. From this initial investigation, other researchers have studied the cytotoxicity by photoexcited TiO_2 on cancer cells [103], but more intriguing has been two recent scientific works where the TiO_2NPs have been found to be effective in PTT against a melanoma cancer model and also as sonosensitizer in SDT against a breast cancer model [104,105].

In the first study, the authors assessed the application of PEGylated TiO_2NPs in inducing hyperthermia and necrosis in in vivo melanoma tumors after PTT consisting of a continuous wave near-infrared (NIR) laser diode at 808 nm wavelength with an intensity of 2 W/cm^2 for seven minutes. Four mice groups were enrolled in the experiments and the main result showed that in the PEGylated TiO_2NPs + laser therapy group, not only did the tumor growth cease, but the tumor size also shrank according to the ultrasonography images and the histopathological examination in the three days following the experiment. Interestingly, five mice from the PEGylated TiO_2NPs + laser therapy group were euthanized after three months of follow-up to demonstrate biocompatibility of these PEGylated TiO_2NPs. However, no data about the survival rate of those animals were reported.

The latter work investigated TiO_2NPs, more precisely the spherical carbon-doped titanium dioxide nanoparticles (C-doped TiO_2NPs), as a sonosensitizer in SDT in order to overcome the major limitation associated with cancer therapies that involve electromagnetic waves, i.e., the shallow penetration depth of light sources into tumor tissue [105]. Taking this into consideration, Yang and colleagues investigated whether C-doped TiO_2NPs were able to suppress the proliferation of 4T1 breast cancer cell line in both in vitro and in vivo models in combination with US treatment (US frequency of 1.0 MHz and a duty cycle of 50% with a negative pressure of 0.33 MPa and intensity of 1.8 W/cm^2 for 90 s) in order to inhibit tumor growth. Firstly, in the in vitro study, the authors quantified ROS production between treatment groups and found that C-doped TiO_2NPs, in combination with US, significantly increased the level of ROS compared to control group. This result corroborated that, under US irradiation, ROS generation could be improved in the presence of C-doped TiO_2NPs. Thereafter SDT cytotoxicity was evaluated confirming that SDT, i.e., 4T1 cells cultured with C-doped TiO_2NPs and subjected to US exposure, induced higher cytotoxicity in 4T1 cells than the other treatment groups. Interesting was the speculation about the possible pathway of cell damage induced by SDT, where the authors suggested a role of sonoluminescence in the C-doped TiO_2NPs activation to generate more ROS and kill 4T1 breast cancer cells [96]. Finally, in this work, to further investigate the cell death of 4T1 cells induced by SDT, an in vivo study was performed. Groups of 5 nude mice were enrolled, all bearing subcutaneous 4TI breast cancer cells, and the data showed that the C-doped TiO_2NPs group (150 mg/mL C-doped TiO_2NPs at day 0 and day 7) and US group (PBS at day 0 and day 7) could not suppress the tumor growth, while the SDT group (150 mg/mL C-doped TiO_2NPs at day 0 and day 7 with US exposure) was able to significantly delay tumor growth in that the relative tumor volume at endpoint was almost half that of other control groups. Moreover, by using histologic staining of the tumor site, authors observed that SDT enhanced the ability to cause 4T1 cell death compared to the other groups, confirming that C-doped TiO_2NPs could be considered as sonosensitizers for sonodynamic treatments, and in general as an efficient strategy for alternative cancer treatments.

2.5. Zinc Oxide NPs

Relying on these promising results from TiO_2NPs as photo- or sonosensitizer in PPT and SDT respectively, ZnONPs have also been investigated as photo- or sonosensitizers for cancer therapy. Indeed, the ZnO electronic structure is part of the semiconducting metal oxide family like TiO_2 and defines ZnONPs' catalytic properties [106]. Therefore, ZnONPs can be photoexcited in the UV-A range ($315 \leq \lambda < 400$ nm) [107]. Another possibility to excite ZnONPs in aqueous solutions is the application of US, as propagation in a liquid milieu causes the well-known physical phenomenon called acoustic cavitation. A number of secondary effects are determined by this phenomenon, and in particular sonoluminescence, which represents a sonoluminescent emission, mainly a UV light, due to the cavitation bubble implosion [96]. For this reason, many researchers suggest that ZnONPs bring about cytotoxic effects when exposed to US via generation of ROS as their principal mechanism of bacterial and cancerous cell death [108].

Through investigating the antibacterial activity of ZnONPs in vitro studies, it is well known that ZnONPs have antimicrobial activity against Gram-positive (*B. cereus, B. subtilis, E. faecium, L. monocytogenes, S. aureus* and *S. epidermis*) and Gram-negative (*E. coli, K. pneumoniae, P. aeruginosa* and *Salmonella sp.*) bacteria [109]. Furthermore, Seil and colleagues have shown that intrinsic antibacterial activity against *S. aureus* of ZnONPs might be enhanced, up to 76%, by US exposure as this provides an additional mechanism to decrease bacterial activity via enhanced generation of hydrogen peroxide by *S. aureus* compared to control samples [110].

Recently Racca et al. [111] reported a novel study demonstrating the highly efficient killing capability of amino-propyl functionalized ZnO nanocrystals (ZnONCs) in association with high-energy shock waves (HESW) for cancer cell treatment. This in vitro work investigated KB cells' viability after incubation with a non-toxic concentration of ZnONCs (10 μg/mL) in combination with HESW exposure. To demonstrate the effective synergy between the HESW and ZnONCs, cells were pre-incubated for 24 h with 10 μg/mL ZnONCs and then exposed to HESW (by considering several energy flux density ranges, for instance 0.15–0.22–0.3-0.4-0.52 mJ/mm^2, corresponding to positive peak pressures of 29.1, 39.4, 50.3, 61.7 and 74.1 MPa, respectively). Authors observed that a single treatment was not able to induce a significant difference in cell viability between HESW exposure alone and ZnONCs + HESW exposure. In contrast, multiple HESW treatments (3 times/day) showed cytotoxicity only for those cells pre-incubated with ZnONCs. Studies of the mechanism showed that the ROS role was controversial. Therefore, the authors suggested that the anticancer activity was due to a combination of several effects, including the non-inertial cavitation, the so- called "nanoscalpel effect," as well as an imbalance of electric change, involving the ZnO piezoelectric behavior [111].

3. Conclusions

Metal-based NPs are highly demanded because they have broad range of applications in healthcare, cosmetics and industry. Particularly, in this review, authors have focused their attention on novel biomedical applications of IONPs, AgNPs, AuNPs, TiO_2NPs and ZnONPs in photodynamic, photothermal and sonodynamic therapies. Specifically, these MNPs are able to be triggered by different physical actuators to generate ROS for the selective killing of bacteria and cancer cells. Their pivotal role as remotely activated NPs for therapeutic ROS generation has been highlighted for IONPs under magnetic field, and for AgNPs, AuNPs, TiO_2NPs and ZnONPs under light or US exposure.

Despite these promising results, MNPs' potential druggability requires further extensive evaluation before they can reach clinical applications. Therefore, future research involving MNPs should consist of robust pre-clinical studies with a predominant focus on acceleration of their clinical translation for biomedical uses.

Author Contributions: Conceptualization, R.C. and L.S.; writing—original draft preparation, R.C., F.F. and L.S.; writing—review and editing, R.C., F.F., T.L. and L.S.; supervision, L.S. All authors have read and agreed to the published version of the manuscript.

Funding: No external funding was received.

Institutional Review Board Statement: Not applicable.

Informed Consent Statement: Not applicable.

Data Availability Statement: No new data were created or analyzed in this study.

Conflicts of Interest: The authors declare no conflict of interest.

References

1. Bayda, S.; Adeel, M.; Tuccinardi, T.; Cordani, M.; Rizzolio, F. The History of Nanoscience and Nanotechnology: From Chemical–Physical Applications to Nanomedicine. *Molecules* **2019**, *25*, 112. [CrossRef] [PubMed]
2. Jeevanandam, J.; Barhoum, A.; Chan, Y.S.; Dufresne, A.; Danquah, M.K. Review on nanoparticles and nanostructured materials: History, sources, toxicity and regulations. *Beilstein J. Nanotechnol.* **2018**, *9*, 1050–1074. [CrossRef] [PubMed]
3. Bruchez, M., Jr.; Moronne, M.; Gin, P.; Weiss, S.; Alivisatos, A.P. Semiconductor Nanocrystals as Fluorescent Biological Labels. *Science* **1998**, *281*, 2013–2016. [CrossRef] [PubMed]
4. Wang, S.; Mamedova, N.; Kotov, N.A.; Chen, W.; Studer, J. Antigen/Antibody Immunocomplex from CdTe Nanoparticle Bioconjugates. *Nano Lett.* **2002**, *2*, 817–822. [CrossRef]
5. Edelstein, R. The BARC biosensor applied to the detection of biological warfare agents. *Biosens. Bioelectron.* **2000**, *14*, 805–813. [CrossRef]
6. Nam, J.-M. Nanoparticle-Based Bio-Bar Codes for the Ultrasensitive Detection of Proteins. *Science* **2003**, *301*, 1884–1886. [CrossRef]
7. Mahtab, R.; Rogers, J.P.; Murphy, C.J. Protein-Sized Quantum Dot Luminescence Can Distinguish between "Straight", "Bent", and "Kinked" Oligonucleotides. *J. Am. Chem. Soc.* **1995**, *117*, 9099–9100. [CrossRef]

8. Ma, J.; Wong, H.; Kong, L.B.; Peng, K.W. Biomimetic processing of nanocrystallite bioactive apatite coating on titanium. *Nanotechnology* **2003**, *14*, 619–623. [CrossRef]
9. Molday, R.S.; Mackenzie, D. Immunospecific ferromagnetic iron-dextran reagents for the labeling and magnetic separation of cells. *J. Immunol. Methods* **1982**, *52*, 353–367. [CrossRef]
10. Shen, Z.; Wu, A.; Chen, X. Iron Oxide Nanoparticle Based Contrast Agents for Magnetic Resonance Imaging. *Mol. Pharm.* **2017**, *14*, 1352–1364. [CrossRef]
11. Thanou, M. Nanoparticles for Drug and Gene Delivery. In *Encyclopedia of Biophysics*; Roberts, G.C.K., Ed.; Springer: Berlin/Heidelberg, Germany, 2013; pp. 1686–1691. ISBN 978-3-642-16711-9.
12. Jin, S.; Ye, K. Nanoparticle-Mediated Drug Delivery and Gene Therapy. *Biotechnol. Prog.* **2007**, *23*, 32–41. [CrossRef] [PubMed]
13. Abrahamse, H.; Kruger, C.A.; Kadanyo, S.; Mishra, A. Nanoparticles for Advanced Photodynamic Therapy of Cancer. *Photomed. Laser Surg.* **2017**, *35*, 581–588. [CrossRef] [PubMed]
14. Abrahamse, H.; Hamblin, M.R. New photosensitizers for photodynamic therapy. *Biochem. J.* **2016**, *473*, 347–364. [CrossRef] [PubMed]
15. Colombeau, L.; Acherar, S.; Baros, F.; Arnoux, P.; Gazzali, A.M.; Zaghdoudi, K.; Toussaint, M.; Vanderesse, R.; Frochot, C. Inorganic Nanoparticles for Photodynamic Therapy. In *Light-Responsive Nanostructured Systems for Applications in Nanomedicine*; Sortino, S., Ed.; Topics in Current Chemistry; Springer International Publishing: Cham, Switzerland, 2016; Volume 370, pp. 113–134, ISBN 978-3-319-22941-6.
16. Zhang, Y.; Yong, L.; Luo, Y.; Ding, X.; Xu, D.; Gao, X.; Yan, S.; Wang, Q.; Luo, J.; Pu, D.; et al. Enhancement of HIFU ablation by sonosensitizer-loading liquid fluorocarbon nanoparticles with pre-targeting in a mouse model. *Sci. Rep.* **2019**, *9*, 6982. [CrossRef]
17. Jaque, D.; Martínez Maestro, L.; del Rosal, B.; Haro-Gonzalez, P.; Benayas, A.; Plaza, J.L.; Martín Rodríguez, E.; García Solé, J. Nanoparticles for photothermal therapies. *Nanoscale* **2014**, *6*, 9494–9530. [CrossRef]
18. Bosca, F.; Foglietta, F.; Gimenez, A.; Canaparo, R.; Durando, G.; Andreana, I.; Barge, A.; Peira, E.; Arpicco, S.; Serpe, L.; et al. Exploiting Lipid and Polymer Nanocarriers to Improve the Anticancer Sonodynamic Activity of Chlorophyll. *Pharmaceutics* **2020**, *12*, 605. [CrossRef]
19. Brazzale, C.; Canaparo, R.; Racca, L.; Foglietta, F.; Durando, G.; Fantozzi, R.; Caliceti, P.; Salmaso, S.; Serpe, L. Enhanced selective sonosensitizing efficacy of ultrasound-based anticancer treatment by targeted gold nanoparticles. *Nanomedicine (London)* **2016**, *11*, 3053–3070. [CrossRef] [PubMed]
20. Foglietta, F.; Canaparo, R.; Francovich, A.; Arena, F.; Civera, S.; Cravotto, G.; Frairia, R.; Serpe, L. Sonodynamic treatment as an innovative bimodal anticancer approach: Shock wave-mediated tumor growth inhibition in a syngeneic breast cancer model. *Discov. Med.* **2015**, *20*, 197–205.
21. Varchi, G.; Foglietta, F.; Canaparo, R.; Ballestri, M.; Arena, F.; Sotgiu, G.; Guerrini, A.; Nanni, C.; Cicoria, G.; Cravotto, G.; et al. Engineered porphyrin loaded core-shell nanoparticles for selective sonodynamic anticancer treatment. *Nanomedicine (London)* **2015**, *10*, 3483–3494. [CrossRef] [PubMed]
22. Abdal Dayem, A.; Hossain, M.; Lee, S.; Kim, K.; Saha, S.; Yang, G.-M.; Choi, H.; Cho, S.-G. The Role of Reactive Oxygen Species (ROS) in the Biological Activities of Metallic Nanoparticles. *Int. J. Mol. Sci.* **2017**, *18*, 120. [CrossRef] [PubMed]
23. Ciccarese, F.; Raimondi, V.; Sharova, E.; Silic-Benussi, M.; Ciminale, V. Nanoparticles as Tools to Target Redox Homeostasis in Cancer Cells. *Antioxidants* **2020**, *9*, 211. [CrossRef] [PubMed]
24. Manke, A.; Wang, L.; Rojanasakul, Y. Mechanisms of Nanoparticle-Induced Oxidative Stress and Toxicity. *BioMed Res. Int.* **2013**, *2013*, 1–15. [CrossRef] [PubMed]
25. Vallyathan, V.; Shi, X. The role of oxygen free radicals in occupational and environmental lung diseases. *Environ. Health Perspect.* **1997**, *105*, 165–177. [CrossRef] [PubMed]
26. Thannickal, V.J.; Fanburg, B.L. Reactive oxygen species in cell signaling. *Am. J. Physiol. Lung Cell. Mol. Physiol.* **2000**, *279*, L1005–L1028. [CrossRef] [PubMed]
27. Maryanovich, M.; Gross, A. A ROS rheostat for cell fate regulation. *Trends Cell Biol.* **2013**, *23*, 129–134. [CrossRef]
28. Sena, L.A.; Chandel, N.S. Physiological Roles of Mitochondrial Reactive Oxygen Species. *Mol. Cell* **2012**, *48*, 158–167. [CrossRef]
29. Cremers, C.M.; Jakob, U. Oxidant Sensing by Reversible Disulfide Bond Formation. *J. Biol. Chem.* **2013**, *288*, 26489–26496. [CrossRef]
30. Snezhkina, A.V.; Kudryavtseva, A.V.; Kardymon, O.L.; Savvateeva, M.V.; Melnikova, N.V.; Krasnov, G.S.; Dmitriev, A.A. ROS Generation and Antioxidant Defense Systems in Normal and Malignant Cells. *Oxidative Med. Cell. Longev.* **2019**, *2019*, 1–17. [CrossRef]
31. Riley, P.A. Free Radicals in Biology: Oxidative Stress and the Effects of Ionizing Radiation. *Int. J. Radiat. Biol.* **1994**, *65*, 27–33. [CrossRef]
32. Poljsak, B.; Šuput, D.; Milisav, I. Achieving the Balance between ROS and Antioxidants: When to Use the Synthetic Antioxidants. *Oxidative Med. Cell. Longev.* **2013**, *2013*, 1–11. [CrossRef]
33. Birben, E.; Sahiner, U.M.; Sackesen, C.; Erzurum, S.; Kalayci, O. Oxidative Stress and Antioxidant Defense. *World Allergy Organ. J.* **2012**, *5*, 9–19. [CrossRef] [PubMed]
34. Forrester, S.J.; Kikuchi, D.S.; Hernandes, M.S.; Xu, Q.; Griendling, K.K. Reactive Oxygen Species in Metabolic and Inflammatory Signaling. *Circ. Res.* **2018**, *122*, 877–902. [CrossRef] [PubMed]
35. Trachootham, D.; Lu, W.; Ogasawara, M.A.; Valle, N.R.-D.; Huang, P. Redox Regulation of Cell Survival. *Antioxid. Redox Signal.* **2008**, *10*, 1343–1374. [CrossRef] [PubMed]
36. Bae, Y.S.; Oh, H.; Rhee, S.G.; Yoo, Y.D. Regulation of reactive oxygen species generation in cell signaling. *Mol. Cells* **2011**, *32*, 491–509. [CrossRef]

37. Valko, M.; Rhodes, C.J.; Moncol, J.; Izakovic, M.; Mazur, M. Free radicals, metals and antioxidants in oxidative stress-induced cancer. *Chem. Biol. Interact.* **2006**, *160*, 1–40. [CrossRef]
38. Bonner, J.C. Lung Fibrotic Responses to Particle Exposure. *Toxicol. Pathol.* **2007**, *35*, 148–153. [CrossRef]
39. Risom, L.; Møller, P.; Loft, S. Oxidative stress-induced DNA damage by particulate air pollution. *Mutat Res.* **2005**, *592*, 119–137. [CrossRef]
40. Li, J.J.; Muralikrishnan, S.; Ng, C.-T.; Yung, L.-Y.L.; Bay, B.-H. Nanoparticle-induced pulmonary toxicity. *Exp. Biol. Med. (Maywood)* **2010**, *235*, 1025–1033. [CrossRef]
41. Huang, C.-C.; Aronstam, R.S.; Chen, D.-R.; Huang, Y.-W. Oxidative stress, calcium homeostasis, and altered gene expression in human lung epithelial cells exposed to ZnO nanoparticles. *Toxicol. Vitr.* **2010**, *24*, 45–55. [CrossRef]
42. Knaapen, A.M.; Borm, P.J.A.; Albrecht, C.; Schins, R.P.F. Inhaled particles and lung cancer. *Part A Mech. Int. J. Cancer* **2004**, *109*, 799–809. [CrossRef]
43. Schins, R.P.F. Mechanisms of genotoxicity of particles and fibers. *Inhal. Toxicol.* **2002**, *14*, 57–78. [CrossRef] [PubMed]
44. Donaldson, K.; Tran, C.L. Inflammation caused by particles and fibers. *Inhal. Toxicol.* **2002**, *14*, 5–27. [CrossRef] [PubMed]
45. Oberdörster, G.; Maynard, A.; Donaldson, K.; Castranova, V.; Fitzpatrick, J.; Ausman, K.; Carter, J.; Karn, B.; Kreyling, W.; Lai, D.; et al. Principles for characterizing the potential human health effects from exposure to nanomaterials: Elements of a screening strategy. *Part. Fibre Toxicol.* **2005**, *2*, 8. [CrossRef] [PubMed]
46. Nel, A. Toxic Potential of Materials at the Nanolevel. *Science* **2006**, *311*, 622–627. [CrossRef]
47. Fubini, B.; Hubbard, A. Reactive oxygen species (ROS) and reactive nitrogen species (RNS) generation by silica in inflammation and fibrosis. *Free Radic. Biol. Med.* **2003**, *34*, 1507–1516. [CrossRef]
48. Smith, K.R.; Klei, L.R.; Barchowsky, A. Arsenite stimulates plasma membrane NADPH oxidase in vascular endothelial cells. *Am. J. Physiol. Lung Cell. Mol. Physiol.* **2001**, *280*, L442–L449. [CrossRef]
49. Sioutas, C.; Delfino, R.J.; Singh, M. Exposure Assessment for Atmospheric Ultrafine Particles (UFPs) and Implications in Epidemiologic Research. *Environ. Health Perspect.* **2005**, *113*, 947–955. [CrossRef]
50. Xia, T.; Kovochich, M.; Brant, J.; Hotze, M.; Sempf, J.; Oberley, T.; Sioutas, C.; Yeh, J.I.; Wiesner, M.R.; Nel, A.E. Comparison of the Abilities of Ambient and Manufactured Nanoparticles To Induce Cellular Toxicity According to an Oxidative Stress Paradigm. *Nano Lett.* **2006**, *6*, 1794–1807. [CrossRef]
51. Hsin, Y.-H.; Chen, C.-F.; Huang, S.; Shih, T.-S.; Lai, P.-S.; Chueh, P.J. The apoptotic effect of nanosilver is mediated by a ROS- and JNK-dependent mechanism involving the mitochondrial pathway in NIH3T3 cells. *Toxicol. Lett.* **2008**, *179*, 130–139. [CrossRef]
52. Eom, H.-J.; Choi, J. p38 MAPK Activation, DNA Damage, Cell Cycle Arrest and Apoptosis As Mechanisms of Toxicity of Silver Nanoparticles in Jurkat T Cells. *Environ. Sci. Technol.* **2010**, *44*, 8337–8342. [CrossRef]
53. Lenaz, G. The Mitochondrial Production of Reactive Oxygen Species: Mechanisms and Implications in Human Pathology. *IUBMB Life (Int. Union Biochem. Mol. Biol. Life)* **2001**, *52*, 159–164. [CrossRef] [PubMed]
54. Turrens, J.F. Mitochondrial formation of reactive oxygen species. *J. Physiol.* **2003**, *552*, 335–344. [CrossRef] [PubMed]
55. Boonstra, J.; Post, J.A. Molecular events associated with reactive oxygen species and cell cycle progression in mammalian cells. *Gene* **2004**, *337*, 1–13. [CrossRef] [PubMed]
56. Shi, Y.; Wang, F.; He, J.; Yadav, S.; Wang, H. Titanium dioxide nanoparticles cause apoptosis in BEAS-2B cells through the caspase 8/t-Bid-independent mitochondrial pathway. *Toxicol. Lett.* **2010**, *196*, 21–27. [CrossRef] [PubMed]
57. Manna, P.; Ghosh, M.; Ghosh, J.; Das, J.; Sil, P.C. Contribution of nano-copper particles to in vivo liver dysfunction and cellular damage: Role of IκBα/NF-κB, MAPKs and mitochondrial signal. *Nanotoxicology* **2012**, *6*, 1–21. [CrossRef]
58. Zhang, X.Q.; Yin, L.H.; Tang, M.; Pu, Y.P. ZnO, TiO(2), SiO(2,) and Al(2)O(3) nanoparticles-induced toxic effects on human fetal lung fibroblasts. *Biomed. Environ. Sci.* **2011**, *24*, 661–669. [CrossRef]
59. Naqvi, S.; Samim, M.; Abdin, M.Z.; Ahmad, F.J.; prashant, C.K.; Dinda, A. Concentration-dependent toxicity of iron oxide nanoparticles mediated by increased oxidative stress. *IJN* **2010**, *2010*, 983–989. [CrossRef]
60. Pujalté, I.; Passagne, I.; Brouillaud, B.; Tréguer, M.; Durand, E.; Ohayon-Courtès, C.; L'Azou, B. Cytotoxicity and oxidative stress induced by different metallic nanoparticles on human kidney cells. *Part. Fibre Toxicol.* **2011**, *8*, 10. [CrossRef]
61. Jeng, H.A.; Swanson, J. Toxicity of Metal Oxide Nanoparticles in Mammalian Cells. *J. Environ. Sci. Health Part A* **2006**, *41*, 2699–2711. [CrossRef]
62. Khalid, K.; Tan, X.; Mohd Zaid, H.F.; Tao, Y.; Lye Chew, C.; Chu, D.-T.; Lam, M.K.; Ho, Y.-C.; Lim, J.W.; Chin Wei, L. Advanced in developmental organic and inorganic nanomaterial: A review. *Bioengineered* **2020**, *11*, 328–355. [CrossRef]
63. Torres-Lugo, M.; Rinaldi, C. Thermal potentiation of chemotherapy by magnetic nanoparticles. *Nanomedicine* **2013**, *8*, 1689–1707. [CrossRef] [PubMed]
64. Spaldin, N.A. *Magnetic Materials: Fundamentals and Applications*, 2nd ed.; Cambridge University Press: Cambridge, UK, 2010; ISBN 978-0-511-78159-9.
65. Russel, W.B.; Saville, D.A.; Schowalter, W.R. *Colloidal Dispersions*, 1st ed.; Cambridge University Press: Cambridge, UK, 1989; ISBN 978-0-521-42600-8.
66. Na, H.B.; Song, I.C.; Hyeon, T. Inorganic Nanoparticles for MRI Contrast Agents. *Adv. Mater.* **2009**, *21*, 2133–2148. [CrossRef]
67. Arruebo, M.; Fernández-Pacheco, R.; Ibarra, M.R.; Santamaría, J. Magnetic nanoparticles for drug delivery. *Nano Today* **2007**, *2*, 22–32. [CrossRef]
68. Dobson Magnetic nanoparticles for gene and drug delivery. *IJN* **2008**, *2008*, 169–180. [CrossRef]

69. Mannix, R.J.; Kumar, S.; Cassiola, F.; Montoya-Zavala, M.; Feinstein, E.; Prentiss, M.; Ingber, D.E. Nanomagnetic actuation of receptor-mediated signal transduction. *Nat. Nanotechnol.* **2008**, *3*, 36–40. [CrossRef]
70. Hua, X.; Yang, Q.; Dong, Z.; Zhang, J.; Zhang, W.; Wang, Q.; Tan, S.; Smyth, H.D.C. Magnetically triggered drug release from nanoparticles and its applications in anti-tumor treatment. *Drug Deliv.* **2017**, *24*, 511–518. [CrossRef]
71. Hervault, A.; Thanh, N.T.K. Magnetic nanoparticle-based therapeutic agents for thermo-chemotherapy treatment of cancer. *Nanoscale* **2014**, *6*, 11553–11573. [CrossRef]
72. Wydra, R.J.; Oliver, C.E.; Anderson, K.W.; Dziubla, T.D.; Hilt, J.Z. Accelerated generation of free radicals by iron oxide nanoparticles in the presence of an alternating magnetic field. *RSC Adv.* **2015**, *5*, 18888–18893. [CrossRef]
73. Wydra, R.J.; Rychahou, P.G.; Evers, B.M.; Anderson, K.W.; Dziubla, T.D.; Hilt, J.Z. The role of ROS generation from magnetic nanoparticles in an alternating magnetic field on cytotoxicity. *Acta Biomater.* **2015**, *25*, 284–290. [CrossRef]
74. Ebrahimi Fard, A.; Zarepour, A.; Zarrabi, A.; Shanei, A.; Salehi, H. Synergistic effect of the combination of triethylene-glycol modified Fe3O4 nanoparticles and ultrasound wave on MCF-7 cells. *J. Magn. Magn. Mater.* **2015**, *394*, 44–49. [CrossRef]
75. Lee, S.; Jun, B.-H. Silver Nanoparticles: Synthesis and Application for Nanomedicine. *Int. J. Mol. Sci.* **2019**, *20*, 865. [CrossRef] [PubMed]
76. Feng, Q.L.; Wu, J.; Chen, G.Q.; Cui, F.Z.; Kim, T.N.; Kim, J.O. A mechanistic study of the antibacterial effect of silver ions on Escherichia coli and Staphylococcus aureus. *J. Biomed. Mater. Res.* **2000**, *52*, 662–668. [CrossRef]
77. Chugh, H.; Sood, D.; Chandra, I.; Tomar, V.; Dhawan, G.; Chandra, R. Role of gold and silver nanoparticles in cancer nano-medicine. *Artif. Cells Nanomed. Biotechnol.* **2018**, *46*, 1210–1220. [CrossRef] [PubMed]
78. Azizi, M.; Ghourchian, H.; Yazdian, F.; Bagherifam, S.; Bekhradnia, S.; Nyström, B. Anti-cancerous effect of albumin coated silver nanoparticles on MDA-MB 231 human breast cancer cell line. *Sci. Rep.* **2017**, *7*, 5178. [CrossRef]
79. Shahverdi, A.R.; Fakhimi, A.; Shahverdi, H.R.; Minaian, S. Synthesis and effect of silver nanoparticles on the antibacterial activity of different antibiotics against Staphylococcus aureus and Escherichia coli. *Nanomed. Nanotechnol. Biol. Med.* **2007**, *3*, 168–171. [CrossRef]
80. da Silva, R.T.P.; Petri, M.V.; Valencia, E.Y.; Camargo, P.H.C.; de Torresi, S.I.C.; Spira, B. Visible Light Plasmon Excitation of Silver Nanoparticles Against Antibiotic-Resistant Pseudomonas aeruginosa. *Microbiology* **2020**, *31*. [CrossRef]
81. Petryayeva, E.; Krull, U.J. Localized surface plasmon resonance: Nanostructures, bioassays and biosensing—A review. *Anal. Chim. Acta* **2011**, *706*, 8–24. [CrossRef]
82. Vasil'kov, A.; Dovnar, R.; Smotryn, S.; Iaskevich, N.; Naumkin, A. Plasmon Resonance of Silver Nanoparticles as a Method of Increasing Their Antibacterial Action. *Antibiotics* **2018**, *7*, 80. [CrossRef]
83. Beer, C.; Foldbjerg, R.; Hayashi, Y.; Sutherland, D.S.; Autrup, H. Toxicity of silver nanoparticles—Nanoparticle or silver ion? *Toxicol. Lett.* **2012**, *208*, 286–292. [CrossRef]
84. Choi, O.; Deng, K.K.; Kim, N.-J.; Ross, L.; Surampalli, R.Y.; Hu, Z. The inhibitory effects of silver nanoparticles, silver ions, and silver chloride colloids on microbial growth. *Water Res.* **2008**, *42*, 3066–3074. [CrossRef]
85. Lee, W.; Kim, K.-J.; Lee, D.G. A novel mechanism for the antibacterial effect of silver nanoparticles on Escherichia coli. *Biometals* **2014**, *27*, 1191–1201. [CrossRef] [PubMed]
86. Jeyaraj, M.; Sathishkumar, G.; Sivanandhan, G.; MubarakAli, D.; Rajesh, M.; Arun, R.; Kapildev, G.; Manickavasagam, M.; Thajuddin, N.; Premkumar, K.; et al. Biogenic silver nanoparticles for cancer treatment: An experimental report. *Colloids Surf. B Biointerfaces* **2013**, *106*, 86–92. [CrossRef] [PubMed]
87. Vasanth, K.; Ilango, K.; MohanKumar, R.; Agrawal, A.; Dubey, G.P. Anticancer activity of Moringa oleifera mediated silver nanoparticles on human cervical carcinoma cells by apoptosis induction. *Colloids Surf. B Biointerfaces* **2014**, *117*, 354–359. [CrossRef] [PubMed]
88. Wang, H.-H.; Su, C.-H.; Wu, Y.-J.; Lin, C.-A.J.; Lee, C.-H.; Shen, J.-L.; Chan, W.-H.; Chang, W.H.; Yeh, H.-I. Application of Gold in Biomedicine: Past, Present and Future. *Int. J. Gerontol.* **2012**, *6*, 1–4. [CrossRef]
89. Bhattacharyya, S.; Kudgus, R.A.; Bhattacharya, R.; Mukherjee, P. Inorganic Nanoparticles in Cancer Therapy. *Pharm. Res.* **2011**, *28*, 237–259. [CrossRef]
90. Yang, W.; Liang, H.; Ma, S.; Wang, D.; Huang, J. Gold nanoparticle based photothermal therapy: Development and application for effective cancer treatment. *Sustain. Mater. Technol.* **2019**, *22*, e00109. [CrossRef]
91. Ash, C.; Dubec, M.; Donne, K.; Bashford, T. Effect of wavelength and beam width on penetration in light-tissue interaction using computational methods. *Lasers Med. Sci.* **2017**, *32*, 1909–1918. [CrossRef]
92. Mendes, R.; Pedrosa, P.; Lima, J.C.; Fernandes, A.R.; Baptista, P.V. Photothermal enhancement of chemotherapy in breast cancer by visible irradiation of Gold Nanoparticles. *Sci. Rep.* **2017**, *7*, 10872. [CrossRef]
93. Abadeer, N.S.; Murphy, C.J. Recent Progress in Cancer Thermal Therapy Using Gold Nanoparticles. *J. Phys. Chem. C* **2016**, *120*, 4691–4716. [CrossRef]
94. Tachibana, K.; Feril, L.B.; Ikeda-Dantsuji, Y. Sonodynamic therapy. *Ultrasonics* **2008**, *48*, 253–259. [CrossRef]
95. Misík, V.; Riesz, P. Free radical intermediates in sonodynamic therapy. *Ann. N. Y. Acad. Sci.* **2000**, *899*, 335–348. [CrossRef] [PubMed]
96. Canaparo, R.; Foglietta, F.; Giuntini, F.; Francovich, A.; Serpe, L. The bright side of sound: Perspectives on the biomedical application of sonoluminescence. *Photochem. Photobiol. Sci.* **2020**. [CrossRef] [PubMed]
97. Li, M.; Yin, J.-J.; Wamer, W.G.; Lo, Y.M. Mechanistic characterization of titanium dioxide nanoparticle-induced toxicity using electron spin resonance. *J. Food Drug Anal.* **2014**, 22, 76–85. [CrossRef] [PubMed]

98. Hirakawa, K.; Hirano, T. Singlet Oxygen Generation Photocatalyzed by TiO_2 Particles and Its Contribution to Biomolecule Damage. *Chem. Lett.* **2006**, *35*, 832–833. [CrossRef]
99. Daimon, T.; Nosaka, Y. Formation and Behavior of Singlet Molecular Oxygen in TiO_2 Photocatalysis Studied by Detection of Near-Infrared Phosphorescence. *J. Phys. Chem. C* **2007**, *111*, 4420–4424. [CrossRef]
100. Kangwansupamonkon, W.; Lauruengtana, V.; Surassmo, S.; Ruktanonchai, U. Antibacterial effect of apatite-coated titanium dioxide for textiles applications. *Nanomed. Nanotechnol. Biol. Med.* **2009**, *5*, 240–249. [CrossRef]
101. Foster, H.A.; Ditta, I.B.; Varghese, S.; Steele, A. Photocatalytic disinfection using titanium dioxide: Spectrum and mechanism of antimicrobial activity. *Appl. Microbiol. Biotechnol.* **2011**, *90*, 1847–1868. [CrossRef]
102. Cai, R.; Kubota, Y.; Shuin, T.; Sakai, H.; Hashimoto, K.; Fujishima, A. Induction of cytotoxicity by photoexcited TiO_2 particles. *Cancer Res.* **1992**, *52*, 2346–2348.
103. Kubota, Y.; Shuin, T.; Kawasaki, C.; Hosaka, M.; Kitamura, H.; Cai, R.; Sakai, H.; Hashimoto, K.; Fujishima, A. Photokilling of T-24 human bladder cancer cells with titanium dioxide. *Br. J. Cancer* **1994**, *70*, 1107–1111. [CrossRef]
104. Behnam, M.A.; Emami, F.; Sobhani, Z.; Dehghanian, A.R. The application of titanium dioxide (TiO_2) nanoparticles in the photo-thermal therapy of melanoma cancer model. *Iran. J. Basic Med. Sci.* **2018**. [CrossRef]
105. Yang, C.-C.; Wang, C.-X.; Kuan, C.-Y.; Chi, C.-Y.; Chen, C.-Y.; Lin, Y.-Y.; Chen, G.-S.; Hou, C.-H.; Lin, F.-H. Using C-doped TiO_2 Nanoparticles as a Novel Sonosensitizer for Cancer Treatment. *Antioxidants* **2020**, *9*, 880. [CrossRef] [PubMed]
106. Khan, M.M.; Adil, S.F.; Al-Mayouf, A. Metal oxides as photocatalysts. *J. Saudi Chem. Soc.* **2015**, *19*, 462–464. [CrossRef]
107. Kim, S.W.; An, Y.-J. Effect of ZnO and TiO_2 nanoparticles preilluminated with UVA and UVB light on Escherichia coli and Bacillus subtilis. *Appl. Microbiol. Biotechnol.* **2012**, *95*, 243–253. [CrossRef] [PubMed]
108. Vighetto, V.; Ancona, A.; Racca, L.; Limongi, T.; Troia, A.; Canavese, G.; Cauda, V. The Synergistic Effect of Nanocrystals Combined With Ultrasound in the Generation of Reactive Oxygen Species for Biomedical Applications. *Front. Bioeng. Biotechnol.* **2019**, *7*, 374. [CrossRef]
109. Vijayakumar, S.; Krishnakumar, C.; Arulmozhi, P.; Mahadevan, S.; Parameswari, N. Biosynthesis, characterization and antimicrobial activities of zinc oxide nanoparticles from leaf extract of Glycosmis pentaphylla (Retz.) DC. *Microb. Pathog.* **2018**, *116*, 44–48. [CrossRef]
110. Seil, J.T.; Webster, T.J. Antibacterial effect of zinc oxide nanoparticles combined with ultrasound. *Nanotechnology* **2012**, *23*, 495101. [CrossRef]
111. Racca, L.; Limongi, T.; Vighetto, V.; Dumontel, B.; Ancona, A.; Canta, M.; Canavese, G.; Garino, N.; Cauda, V. Zinc Oxide Nanocrystals and High-Energy Shock Waves: A New Synergy for the Treatment of Cancer Cells. *Front. Bioeng. Biotechnol.* **2020**, *8*, 577. [CrossRef]

MDPI
St. Alban-Anlage 66
4052 Basel
Switzerland
Tel. +41 61 683 77 34
Fax +41 61 302 89 18
www.mdpi.com

Materials Editorial Office
E-mail: materials@mdpi.com
www.mdpi.com/journal/materials

www.ingramcontent.com/pod-product-compliance
Lightning Source LLC
Chambersburg PA
CBHW062103220926
51850CB00010B/2878
* 9 7 8 3 0 3 6 5 3 1 6 9 4 *